OXFORD REVISION G

A Level

Advanced BIOLOGY through diagrams

W R Pickering

Oxford University Press

Oxford University Press, Great Clarendon Street, Oxford OX2 6DP

Oxford New York
Athens Auckland Bogota Bangkok Bombay
Buenos Aires Calcutta Cape Town Dar es Salaam
Delhi Florence Hong Kong Istanbul Karachi
Kuala Lumpur Madras Madrid Melbourne
Mexico City Nairobi Paris Singapore
Taipei Tokyo Toronto Warsaw

and associated companies in
Berlin Ibadan

Oxford is a trade mark of Oxford University Press

First published 1994
Reprinted 1995 (twice), 1996 (with revised index), 1997

New edition 1998 ISBN 0 19 914719 1(School edition)
0 19 914720 5 (Bookshop edition)

Typesetting, design and illustration by Hardlines, Charlbury, Oxford
Printed in Great Britain

CONTENTS

CELL STRUCTURE AND BIOCHEMISTRY

PLANT PHYSIOLOGY

ECOLOGY AND CONSERVATION

ANIMAL PHYSIOLOGY

GENETICS AND GENETIC ENGINEERING

Use of the light microscope

PREPARATION FOR LIGHT MICROSCOPY

FIXATION – preserves material in a life-like condition with minimum distortion

DEHYDRATION – removes traces of water from the fixed material

CLEARING – removes dehydrating alcohol so that material is made transparent

EMBEDDING – supports the material so that it is firm enough for sectioning

SECTIONING – prepares slices of material which are thin enough to allow light to pass through

STAINING – improves contrast between different structures (most biological material is transparent)

MOUNTING – embeds and protects material so that it is suitable for viewing over a long period

Methylene blue
- nuclei stain blue

Leishman's stain
- blood cells stain pink
- white blood cell nuclei stain blue

Safranin/light green
- a plant cell stain
- cytoplasm and cellulose stain green
- nuclei and lignin stain red
- chloroplasts stain pink

Haematoxylin/eosin
- nuclei stain blue
- cytoplasm stains pink

Feulgens stain
- chromosomes during cell division stain purple

Aniline blue
- fungal hyphae and spores stain deep blue

Eyepiece: produces a ***'real image'***;

magnifies but does not *resolve* the image produced by the objective lens;

the eyepiece may be dismantled so that an *eyepiece graticule* may be inserted if the microscope is to be *used for measurement*.

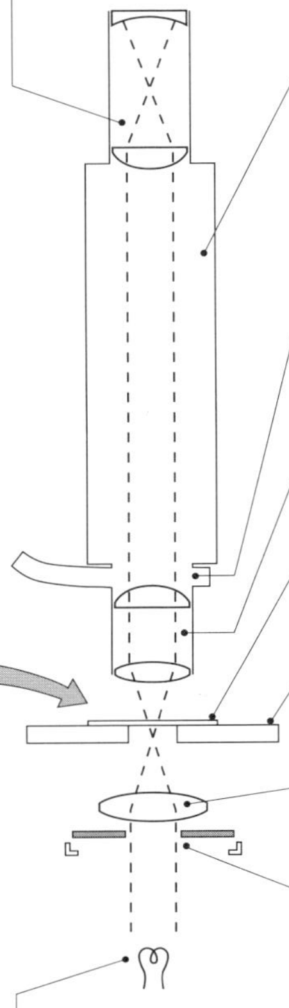

Barrel: route for light rays from objective lens;

may be moved, using a simple racking system, so that object is in focus.

Turret: holds 2, 3 or 4 objective lenses, and can be rotated so that lenses of different focal lengths (hence magnification) can be used.

Objective lens: responsible for both *magnification* and *resolution* of the object.

Specimen/object: is supported on a transparent glass slide.

Stage: holds specimen in correct position relative to optical system at 90° to light path.

Condenser: the condenser focuses the light from the illuminator on to the specimen.

Iris diaphragm: controls amount of light reaching specimen. Best definition is obtained by *reduction* of intensity, not by its increase.

Substage illumination: 'white' light is most commonly used. Light of shorter wavelength (e.g. blue light), produced by changing bulb or with a system of filters, improves *resolution* of the object.

Light must only come from substage position: none on stage.

Transmission electron microscope

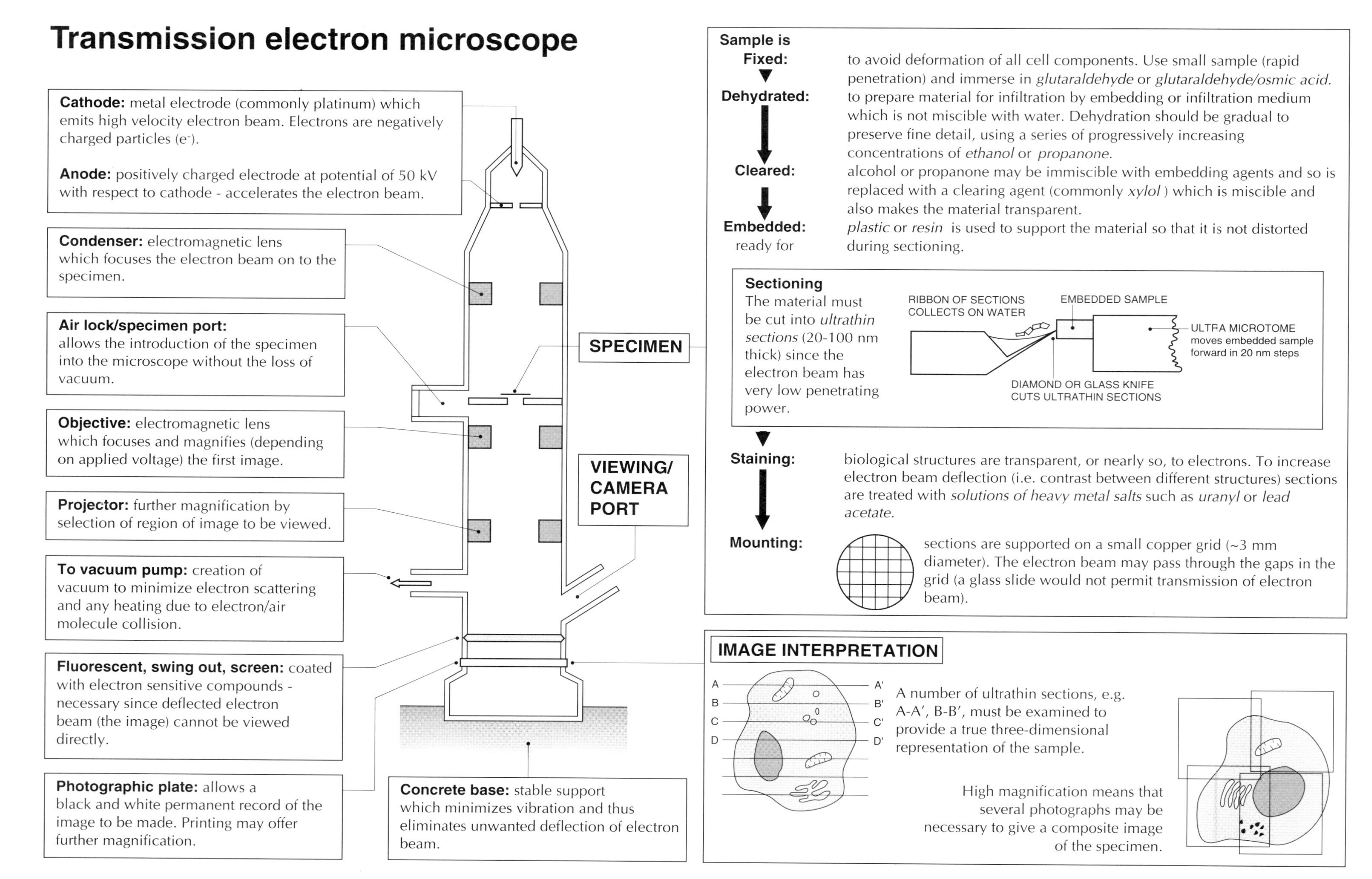

Physical properties of water

are explained by hydrogen bonding between the individual molecules

Solvent properties The polarity of water makes it an excellent solvent for other polar molecules …

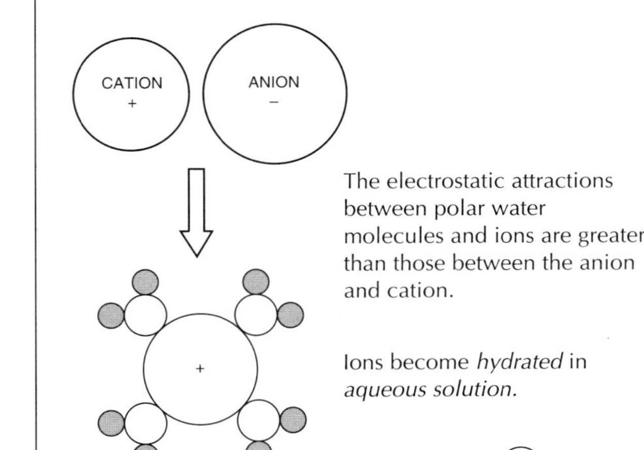

The electrostatic attractions between polar water molecules and ions are greater than those between the anion and cation.

Ions become *hydrated* in *aqueous solution.*

Such polar substances, which dissolve in water, are said to be *hydrophilic* ('water-loving').

… but means that non-polar (*hydrophobic* or 'water-hating') substances do not readily dissolve in water.

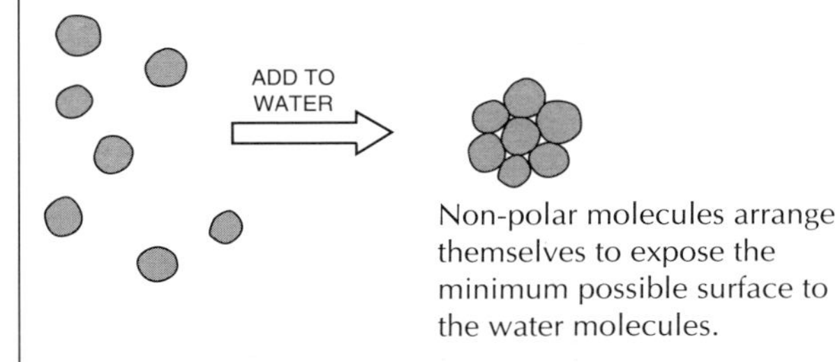

Non-polar molecules arrange themselves to expose the minimum possible surface to the water molecules.

Because hydrogen and oxygen atoms are different in *size* and *electronegativity* the water molecule (H_2O) is *non-linear* and *polar.*

Hydrogen bond - one water molecule may form hydrogen bonds with up to ***four*** other water molecules.

This polarity means that individual water molecules can form *hydrogen bonds* with other water molecules. Although these individual hydrogen bonds are weak, collectively *they make water a much more stable substance than would otherwise be the case.*

High specific heat capacity The specific heat capacity of water (the amount of heat, measured in joules, required to raise 1 kg of water through 1°C) is very high: much of the heat absorbed is used to break the hydrogen bonds which hold the water molecules together.

High latent heat of vaporization Hydrogen bonds attract molecules of liquid water to one another and make it difficult for the molecules to escape as vapour: thus a relatively high energy input is necessary to vaporize water and water has a much higher boiling point than other molecules of the same size.

Molecular mobility The weakness of individual hydrogen bonds means that individual water molecules continually jostle one another when in the liquid phase.

Cohesion and surface tension Hydrogen bonding causes water molecules to 'stick together', and also to stick to other molecules - the phenomenon of ***cohesion***. At the surface of a liquid the inwardly-acting cohesive forces produce a 'surface tension' as the molecules are particularly attracted to one another.

Density and freezing properties As water cools towards its freezing point the individual molecules slow down sufficiently for each one to form its maximum number of hydrogen bonds. To do this the water molecules in liquid water must move further apart to give enough space for all four hydrogen bonds to fit into. As a result water expands as it freezes, so that ice is less dense than liquid water and therefore floats upon its surface.

Colloid formation Some molecules have strong intramolecular forces which prevent their solution in water, but have charged surfaces which attract a covering of water molecules. This covering ensures that the molecules remain dispersed throughout the water, rather than forming large aggregates which could settle out. The dispersed particles and the liquid around them collectively form a ***colloid***.

The biological importance of water depends on its physical properties

Solvent properties:
allow water to act as a transport medium for polar solutes.
For example,
movements of minerals to lakes and seas;
transport via blood and lymph in multicellular animals;
removal of metabolic wastes such as urea and ammonia in urine.

Transpiration stream: the continuous column of water is able to move up the xylem because of cohesion between water molecules and adhesion between water and the walls of the xylem vessels.

Molecular mobility: the rather weak nature of individual hydrogen bonds means that water molecules can move easily relative to one another - this allows *osmosis* (vital for uptake and movement of water) to take place.

Expansion on freezing: since ice floats it forms at the surface of ponds and lakes - it therefore insulates organisms in the water below it, and allows the ice to thaw rapidly when temperatures rise. Changes in density also maintain circulation in large bodies of water, thus helping nutrient cycling. Floating ice also means that penguins and polar bears have somewhere to stand!

Metabolic functions
Water is used directly …
1. as a reagent (source of reducing power) in photosynthesis
2. to hydrolyse macromolecules to their subunits, in digestion for example.

… and is also the medium in which all biochemical reactions take place.

Volatility/stability: is balanced at Earth's temperatures so that a water cycle of evaporation, transpiration and precipitation is maintained.

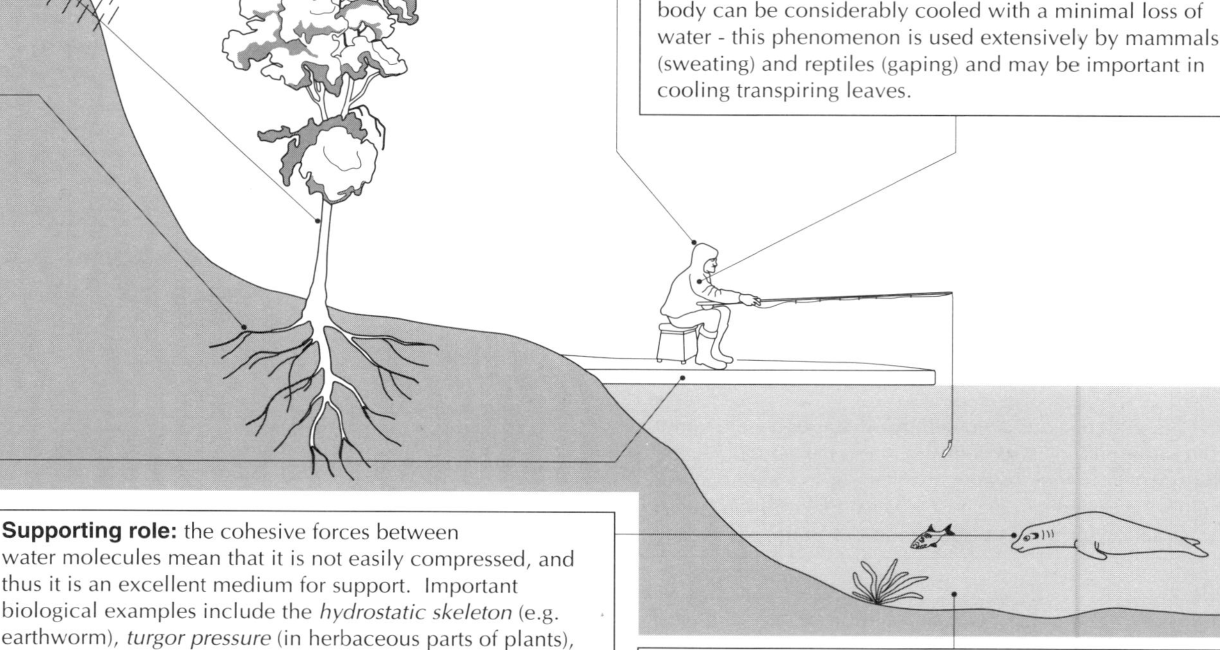

Supporting role: the cohesive forces between water molecules mean that it is not easily compressed, and thus it is an excellent medium for support. Important biological examples include the *hydrostatic skeleton* (e.g. earthworm), *turgor pressure* (in herbaceous parts of plants), *amniotic fluid* (which supports and protects the mammalian foetus) and as a *general supporting medium* (particularly for large aquatic mammals such as whales).

Lubricant properties: water's cohesive and adhesive properties mean that it is viscous, making it a useful lubricant in biological systems.
For example,
synovial fluid - lubricates many vertebrate joints;
pleural fluid - minimizes friction between lungs and thoracic cage (ribs) during breathing;
mucus - permits easy passage of faeces down the colon, and lubricates the penis and vagina during intercourse.

Thermoregulation: the high specific heat capacity of water means that bodies composed largely of water (cells are typically 70-80% water) are very thermostable, and thus less prone to heat damage by changes in environmental temperatures.
The high latent heat of vaporization of water means that a body can be considerably cooled with a minimal loss of water - this phenomenon is used extensively by mammals (sweating) and reptiles (gaping) and may be important in cooling transpiring leaves.

Transparency: water permits the passage of visible light. This means that photosynthesis (and associated food chains) is possible in relatively shallow aquatic environments.

Osmosis

Water molecules, like other molecules, are mobile. In pure water, or in solutions containing very few solute molecules, the water molecules can move very freely (they have a high ***free kinetic energy***). As a result, many of the water molecules may cross the membrane, which is freely permeable to water.

Partially permeable membrane allows the free passage of some particles but is not freely permeable to others. Biological membranes are ***freely permeable to water*** but have ***restricted permeability to solutes*** such as sodium ions and glucose molecules, i.e. they are ***selectively permeable.***

In a solution with many solute molecules the movement of the water molecules is restricted because of solute-water interactions. Fewer of the water molecules have a ***free kinetic energy*** which is great enough to enable them to cross the membrane.

Solute molecules cannot cross the membrane as freely or as rapidly as water molecules can.

MANY WATER MOLECULES CAN MOVE IN THIS DIRECTION

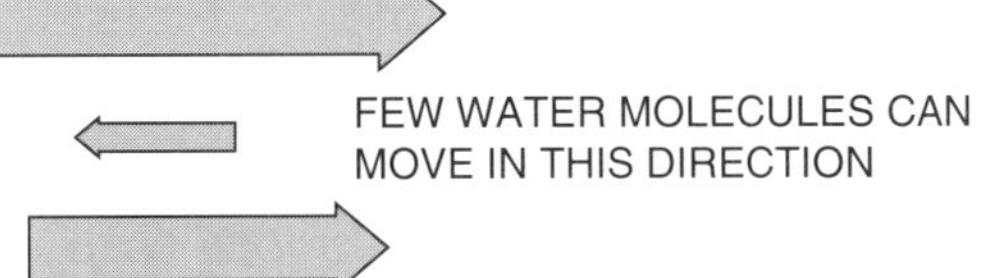

FEW WATER MOLECULES CAN MOVE IN THIS DIRECTION

THERE IS A NET MOVEMENT OF WATER MOLECULES IN THIS DIRECTION *

This movement of water depends on how many water molecules have sufficient free kinetic energy to 'escape from' the system

so that any system in which the water molecules have a ***high*** average kinetic energy will have a greater tendency to lose water than will a system in which the water molecules have a ***low*** average kinetic energy

and when describing water movements scientists replace the term ***free kinetic energy*** with the term ***water potential***, so that

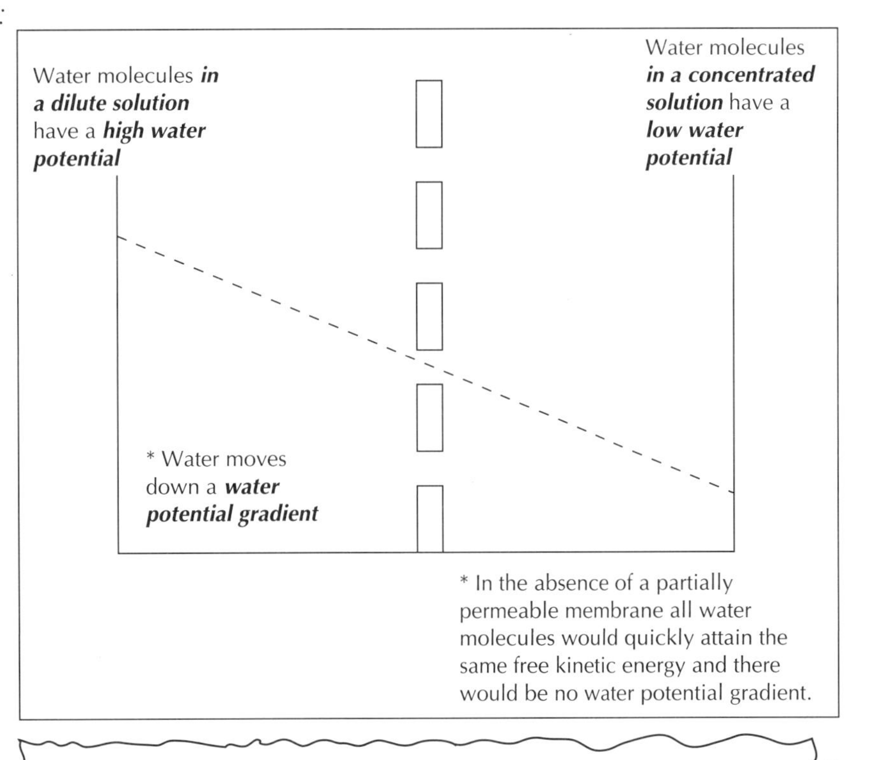

* In the absence of a partially permeable membrane all water molecules would quickly attain the same free kinetic energy and there would be no water potential gradient.

Osmosis is

- *** the movement of water***
- *** down a water potential gradient***
- *** across a partially permeable membrane***
- *** to a solution with a more negative water potential.***

Structural components of membranes

permit fluidity, selective transport and recognition, integrity and compartmentalization.

Because of the different solubility properties of the two ends of phospholipid molecules ...

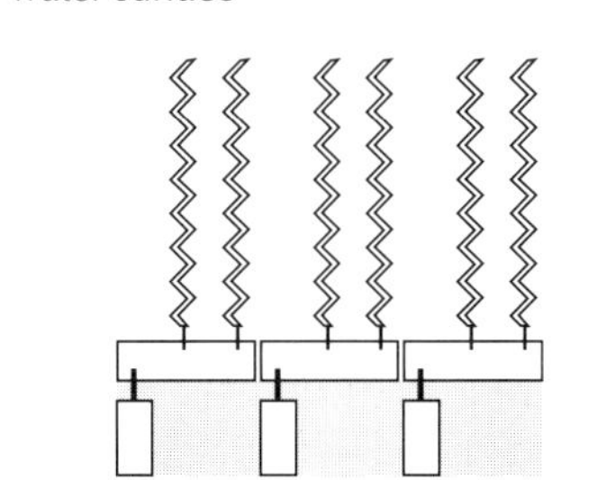

... such molecules form a layer at a water surface

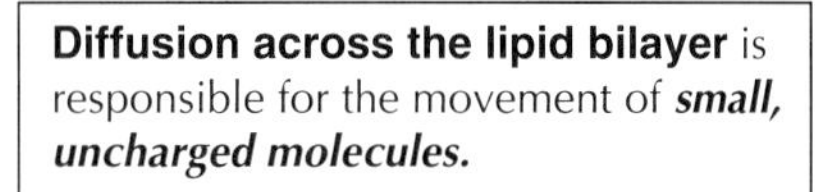

and a ***phospholipid bilayer*** can act as a barrier between two aqueous environments.

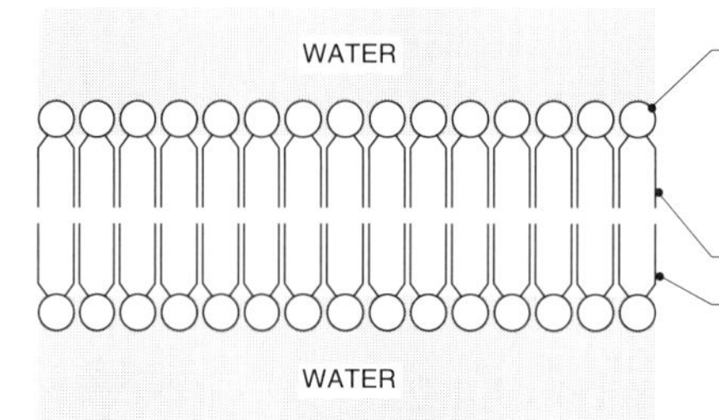

Hydrophilic heads point outwards: form hydrogen bonds with water

Hydrophobic tails point towards one another: this maximizes hydrophobic attractions and excludes water

Lipid composition influences membrane fluidity: unsaturated fatty acid tails are 'kinked', limit close packing of the hydrophobic tails and so ***increase*** fluidity, but cholesterol may interfere with lateral movement of hydrophobic tails and thus ***reduce*** membrane fluidity.

Surface carbohydrates (collectively the ***glycocalyx)*** are usually oligosaccharides which are positioned to aid in cell recognition functions.

Diffusion across the lipid bilayer is responsible for the movement of ***small, uncharged molecules.***

Thus O_2, H_2O, CO_2, urea and ethanol cross rapidly (they 'squeeze between') the polar phospholipid heads then dissolve in the lipid on one side of the membrane and emerge on the other.

Large or ***charged molecules*** cannot cross the lipid bilayer.

Thus Na^+, K^+, Cl^-, HCO_3^- and glucose do not cross in this way.

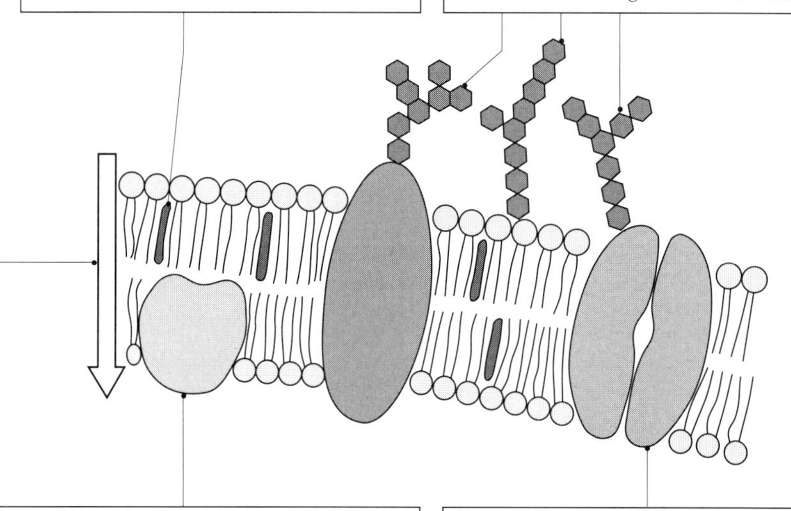

Active transport uses a ***carrier protein*** to transport a solute across a membrane but ***energy is required*** since transport may be ***against a concentration gradient.*** Typically ATP is hydrolysed and the binding of the phosphate group to the carrier changes the protein's conformation in such a way that the solute molecule is moved across the membrane.

Facilitated diffusion uses a ***carrier protein*** to transfer a molecule across a membrane ***along*** its electrochemical gradient. The binding of the solute alters the conformation of the carrier so that its position in the membrane changes and the solute molecule is discharged on the other side of the membrane. Glucose uptake by erythrocytes occurs in this way.
N.B. There is ***no requirement for ATP***, as there is ***no energy consumption.***

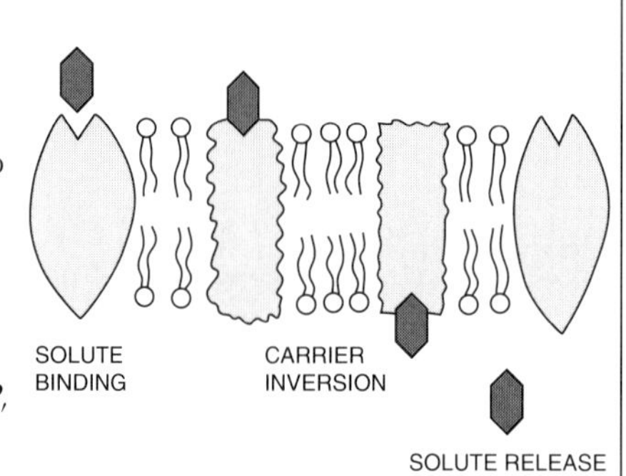

Diffusion through aqueous channels in pore proteins: transmembrane proteins may have aqueous channels through which charged molecules may pass and thus avoid the hydrophobic tails of the phospholipid molecules.

Na^+

Some channels are open all of the time, but others are ***gated*** (they open and close only in response to a stimulus, such as a change in the membrane's electrical potential). Such ***gated channels*** are vital to the operation of nerve and muscle, where movements of Na^+, K^+ and Ca^{2+} initiate information transfer.

Animal cell ultrastructure

Lysosomes are sacs that contain high concentrations of hydrolytic (digestive) enzymes. These enzymes are kept apart from the cell contents which they would otherwise destroy, and they are kept inactive by an alkaline environment within the lysosome. They are especially abundant in cells with a high phagocytic activity, such as some *neutrophils.*

Free ribosomes are the sites of protein synthesis, principally for proteins destined for intracellular use. There may be 50 000 or more in a typical eukaryote cell.

Endocytic vesicle may contain molecules or structures too large to cross the membrane by active transport or diffusion.

Microtubules are hollow tubes of the protein *tubulin,* about 25 nm in diameter. They are involved in intracellular transport (e.g. the movement of mitochondria), have a structural role as part of the cytoskeleton and are components of other specialized structures such as the centrioles and the basal bodies of cilia and flagella.

Nucleus is the centre of the regulation of cell activities since it contains the hereditary material, DNA, carrying the information for protein synthesis. The DNA is bound up with histone protein to form chromatin. The nucleus contains one or more nucleoli in which ribosome subunits, ribosomal RNA, and transfer RNA are manufactured. The nucleus is surrounded by a double nuclear membrane, crossed by a number of nuclear pores. The nucleus is continuous with the endoplasmic reticulum. There is usually only one nucleus per cell, although there may be many in very large cells such as those of striated (skeletal) muscle. Such multinucleate cells are called coenocytes.

Mitochondrion (pl. mitochondria) is the site of aerobic respiration. Mitochondria have a highly folded inner membrane which supports the proteins of the electron transport chain responsible for the synthesis of ATP by oxidative phosphorylation. The mitochondrial matrix contains the enzymes of the TCA cycle, an important metabolic 'hub'. These organelles are abundant in cells which are physically *(skeletal muscle)* and metabolically *(hepatocytes)* active.

Microvilli are extensions of the plasmamembrane which increase the cell surface area. They are commonly abundant in cells with a high absorptive capacity, such as *hepatocytes* or cells of the *first coiled tubule of the nephron.* Collectively the microvilli represent a *brush border* to the cell.

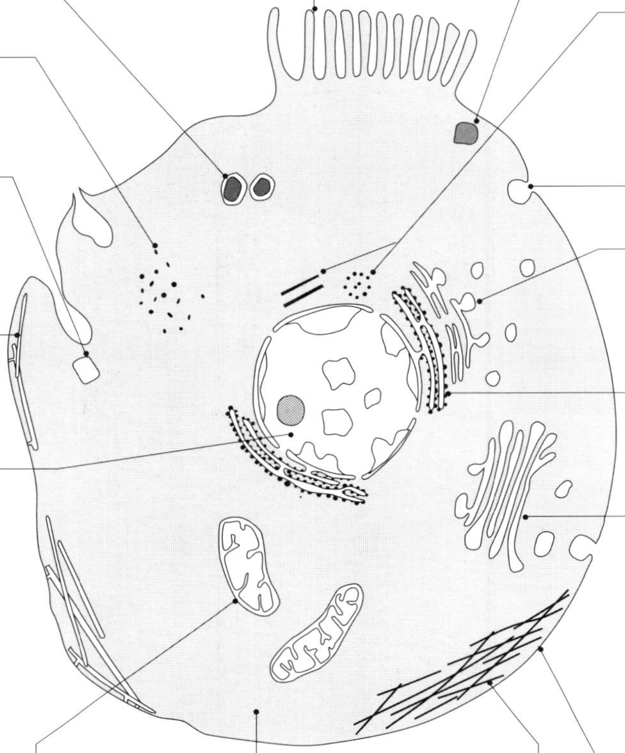

Peroxisome is one of the group of vesicles known as *microbodies.* Each of them contains oxidative enzymes such as *catalase,* and they are particularly important in delaying cell ageing.

Centrioles are a pair of structures, held at right angles to one another, which act as organizers of the nuclear spindle in preparation for the separation of chromosomes or chromatids during nuclear division.

Secretory vesicle undergoing exocytosis. May be carrying a synthetic product of the cell (such as a protein packaged at the Golgi body) or the products of degradation by lysosomes. Secretory vesicles are abundant in cells with a high synthetic activity, such as the cells of the *Islets of Langerhans.*

Smooth endoplasmic reticulum is a series of flattened sacs and sheets that are the sites of synthesis of steroids and lipids.

Rough endoplasmic reticulum is so-called because of the many ribosomes attached to its surface. This intracellular membrane system aids cell compartmentalization and transports proteins synthesized at the ribosomes towards the Golgi bodies for secretory packaging.

Golgi apparatus consists of a stack of sacs called *cisternae.* It modifies a number of cell products delivered to it, often enclosing them in vesicles to be secreted. Such products include trypsinogen (from *pancreatic acinar cells*), insulin (from *beta-cells of the Islets of Langerhans*) and mucin (from *goblet cells in the trachea*). The Golgi is also involved in lipid modification in cells of the ileum, and plays a part in the formation of lysosomes.

Cytoplasm is principally water, with many solutes including glucose, proteins and ions. It is permeated by the *cytoskeleton,* which is the main architectural support of the cell.

Microfilaments are threads of the protein *actin.* They are usually situated in bundles just beneath the cell surface and play a role in endo- and exocytosis, and possibly in cell motility.

Plasmalemma (plasmamembrane) is the surface of the cell and represents its contact with its environment. It is differentially permeable and regulates the movement of solutes between the cell and its environment. There are many specializations of the membrane, often concerning its protein content.

Typical plant cell contains chloroplasts and a permanent vacuole, and is surrounded by a cellulose cell wall.

Chloroplast is the site of photosynthesis. It is one of a number of plastids, all of which develop from *proplastids* which are small, pale green or colourless organelles.

Other typical plastids of complex cells are *chromoplasts* which may develop from chloroplasts by internal rearrangements. Chromoplasts are coloured due to the presence of carotenoid pigments and are most abundant in cells of flower petals or fruit skins.

Leucoplasts are a third type of plastid common in cells of higher plants - they include *amyloplasts* which synthesize and store starches and *elaioplasts* which synthesize oils.

Vacuole may occupy 90% of the volume of a mature plant cell. It is filled with cell sap (a solution of salts, sugars and organic acids) and helps to maintain turgor pressure inside the cell. The vacuole also contains anthocyanins, pigments responsible for many of the red, blue and purple colours of flowers. Vacuoles also contains enzymes involved in recycling of cell components such as chloroplasts. The vacuolar membrane is called the *tonoplast.*

Microtubules are hollow structures (about 25 nm in diameter) composed of the protein tubulin. They occur just below the plasmamembrane where they may aid the addition of cellulose to the cell wall. They are also involved in the cytoplasmic streaming of organelles such as Golgi bodies and chloroplasts, and they form the spindles and cell plates of dividing cells.

Plasmamembrane (plasmalemma, cell surface membrane) is the differentially-permeable cell surface, responsible for the control of solute movements between the cell and its environment. It is flexible enough to move close to or away from the cell wall as the water content of the cytoplasm changes. The membrane is also responsible for the synthesis and assembly of cell wall components.

Golgi body (dictyosome) synthesizes polysaccharides and packages them in vesicles which migrate to the plasmamembrane for eventual incorporation in the cell wall.

Mitochondrion contains the enzyme systems for ATP synthesis by oxidative phosphorylation. May be abundant in sieve tube companion cells, root epidermal cells and dividing meristematic cells.

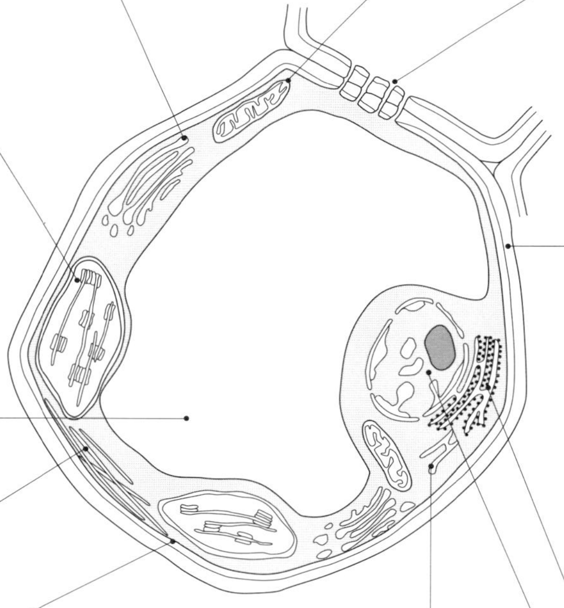

Plasmodesmata are minute strands of cytoplasm which pass through pores in the cell wall and connect the protoplasts of adjacent cells. This represents the *symplast* pathway for the movement of water and solutes throughout the plant body. These cell-cell cytoplasm connections are important in cell survival during periods of drought. The E.R. of adjacent cells is also in contact through these strands.

Cell wall is composed of long cellulose molecules grouped in bundles called *microfibrils* which, in turn, are twisted into rope-like *macrofibrils.* The macrofibrils are embedded in a matrix of *pectins* (which are very adhesive) and *hemicelluloses* (which are quite fluid). There may be a *secondary cell wall,* in which case the outer covering of the cell is arranged as:

Plasmalemma

Secondary cell wall: laid down on inside of primary wall. Often impregnated with *lignin* (gives mechanical strength to xylem) or *suberin* (waterproofs endodermis).

Primary cell wall: laid down first, by plasmamembrane.

Middle lamella: contains gums and calcium pectate to cement cells together.

The function of the cell wall is a mechanical one - pressure from the cell protoplast maintains cell turgidity. The wall is freely permeable to water and most solutes so that the cell wall represents an important transport route - the *apoplast system* - throughout the plant body.

Rough endoplasmic reticulum is the site of protein synthesis (on the attached ribosomes), storage and preparation for secretion. The endoplasmic reticulum (E.R.) also plays a part in the compartmentalization of the cell.

Smooth endoplasmic reticulum is the site of lipid synthesis and secretion.

Nucleus is surrounded by the nuclear envelope and contains the genetic material, DNA, associated with histone protein to form chromatin. The nucleus thus controls the activity of the cell through its regulation of protein synthesis. The nucleolus is the site of synthesis of transfer RNA, ribosomal RNA, and ribosomal subunits.

Cell membrane systems are important in intracellular division of labour. They allow compartmentalization and therefore efficiency through locations of multi-enzyme pathways.

Reverse pinocytosis (exocytosis) releases contents of vesicle into the extracellular environment.

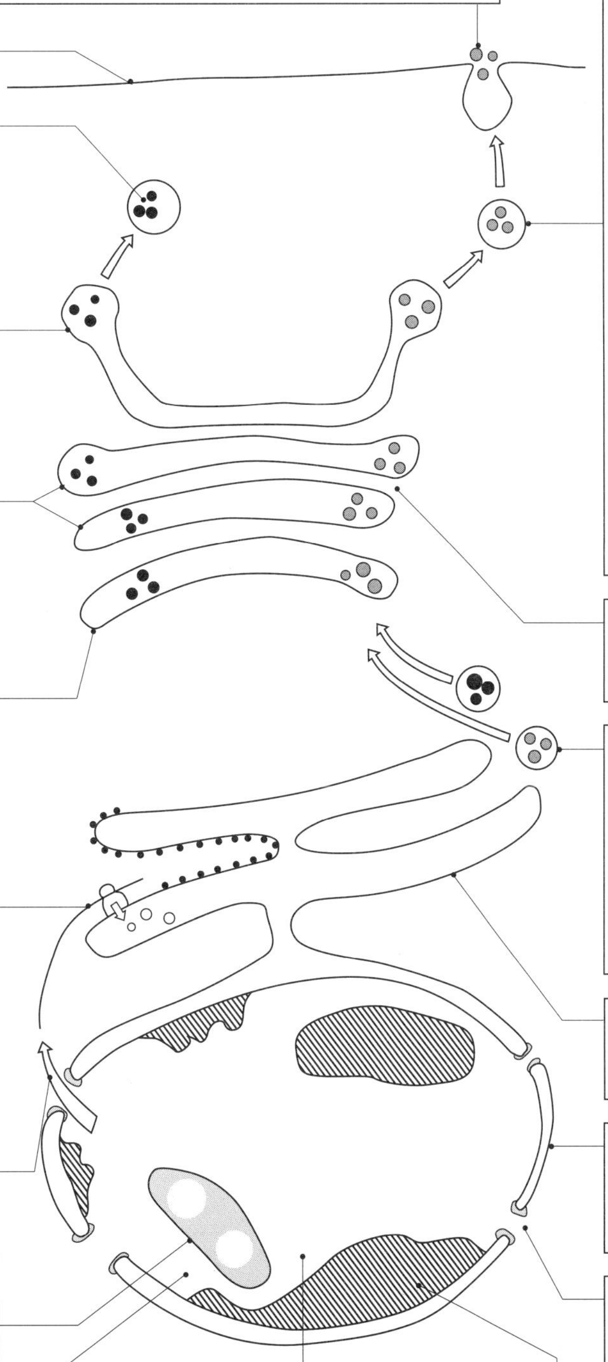

Plasmamembrane

Lysosome contains hydrolytic enzymes which may digest ingested materials, redundant organelles ***(autophagy)*** or whole cells ***(autolysis).***

Trans Golgi network separates products ready for inclusion in secretory vesicles or in lysosomes.

Cisternae stack of Golgi 'processes' molecules often by adding or modifying carbohydrate 'signals' which direct the molecules to the correct cellular compartment.

Cis Golgi network collects sacs from the E.R. Any misdirected molecules, e.g. components of E.R. enzyme systems, are returned to the endoplasmic reticulum.

Protein synthesis at ribosomes on E.R. Newly-synthesized protein carries a 'signal' which ensures that the protein will enter the cisterna ready to be packaged within a sac and delivered to the Golgi apparatus.

Messenger RNA carries coded message for protein synthesis from nucleus to ribosomes.

Nucleolus is the site of manufacture of ribosomal subunits. It disperses in preparation for nuclear division, and is reassembled at the end of telophase.

Nucleoplasm contains a variety of solutes, including nucleoside triphosphates for DNA synthesis, and the enzyme complex (DNA polymerase) which regulates DNA replication and repair.

Exocytic vesicle contains product for export, e.g. mucoprotein from goblet cells, trypsinogen from pancreatic acinar cells and complex carbohydrates for plant cell wall synthesis.

The ***endoplasmic reticulum (E.R.)*** has a wide range of other functions:

1. ***Synthesis of lipids,*** e.g. reassembly of fats in gut epithelium.
2. ***Steroid synthesis,*** e.g. in cells which secrete steroid hormones.
3. ***Control of Ca^{2+} concentration*** in skeletal muscle cells.

1, 2 and 3 all occur on smooth E.R.

4. ***Surface for enzyme systems,*** e.g. the oxidizing system which detoxifies alcohol and other drugs in the liver.

Product molecules are moved through the stack in a precisely defined sequence.

Endoplasmic reticulum 'buds off' membranous sacs containing products of its metabolism. These products include proteins and lipids and may be for export (● grey) or for use within the cell (●).

Cisterna is an enclosed space within the membranes of the E.R.

Nuclear envelope is a double membrane, the outer of which is continuous with the E.R.

Nuclear pore can regulate the entry (e.g. ribosomal proteins) and exit (e.g. ribosomal subunits, messenger RNA) of molecules to and from the nucleus.

Nucleus contains coded information for protein synthesis as a series of genes on the chromosomes.

Chromatin is the genetic material, containing the coded information for protein synthesis in the cell. It is composed of DNA bound to basic proteins called ***histones.*** The DNA and histone are organized into ***nucleosomes.*** During nuclear division the chromatin condenses to form the ***chromosomes,*** and the chromatin containing DNA which is being 'expressed' (transcribed into mRNA) becomes visible as more loosely-coiled threads called ***euchromatin.***

A prokaryotic cell (e.g. a bacterium) has no true organelles.

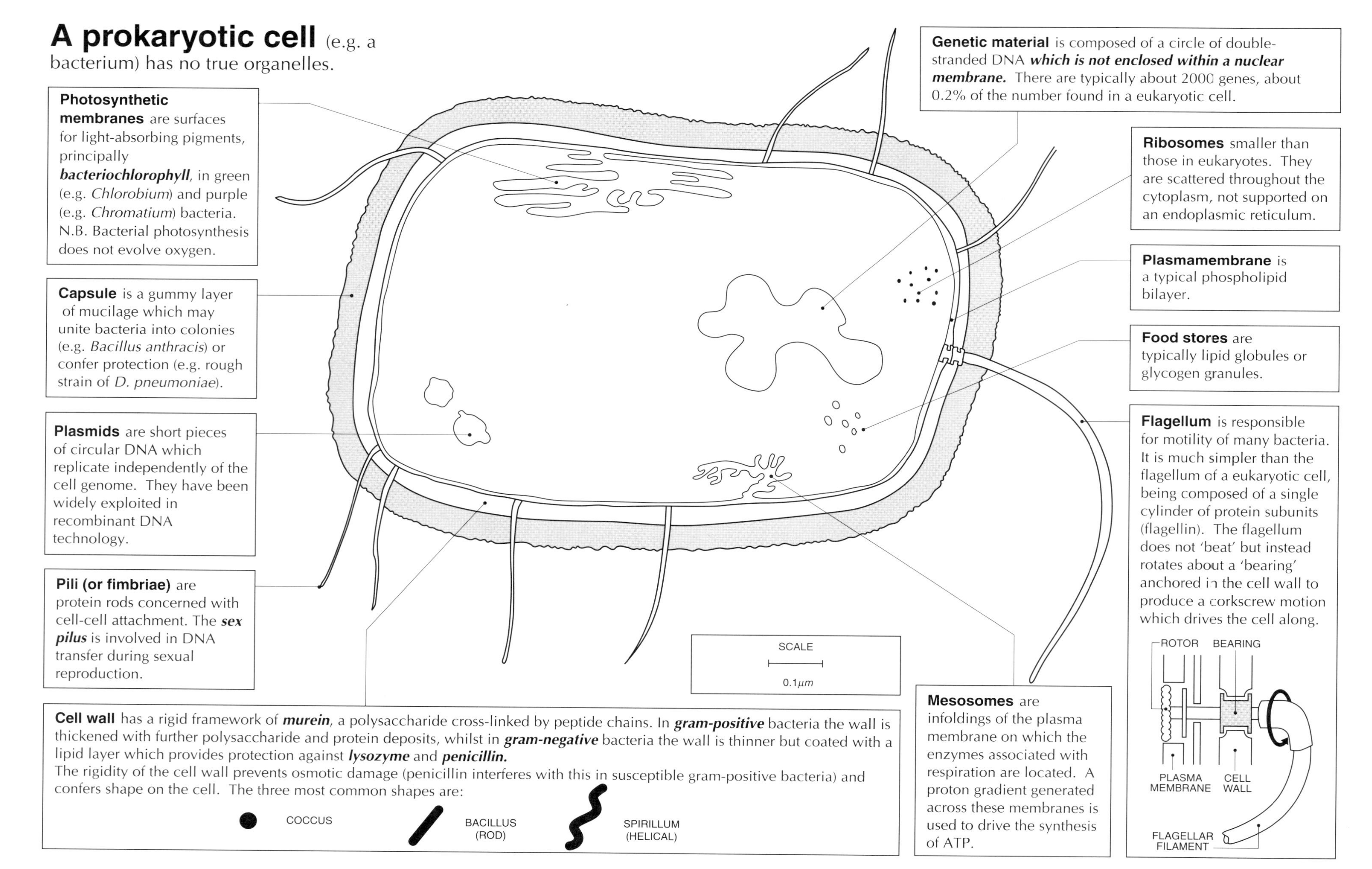

Photosynthetic membranes are surfaces for light-absorbing pigments, principally ***bacteriochlorophyll***, in green (e.g. *Chlorobium*) and purple (e.g. *Chromatium*) bacteria. N.B. Bacterial photosynthesis does not evolve oxygen.

Capsule is a gummy layer of mucilage which may unite bacteria into colonies (e.g. *Bacillus anthracis*) or confer protection (e.g. rough strain of *D. pneumoniae*).

Plasmids are short pieces of circular DNA which replicate independently of the cell genome. They have been widely exploited in recombinant DNA technology.

Pili (or fimbriae) are protein rods concerned with cell-cell attachment. The ***sex pilus*** is involved in DNA transfer during sexual reproduction.

Cell wall has a rigid framework of ***murein***, a polysaccharide cross-linked by peptide chains. In ***gram-positive*** bacteria the wall is thickened with further polysaccharide and protein deposits, whilst in ***gram-negative*** bacteria the wall is thinner but coated with a lipid layer which provides protection against ***lysozyme*** and ***penicillin.***
The rigidity of the cell wall prevents osmotic damage (penicillin interferes with this in susceptible gram-positive bacteria) and confers shape on the cell. The three most common shapes are:

Genetic material is composed of a circle of double-stranded DNA ***which is not enclosed within a nuclear membrane.*** There are typically about 2000 genes, about 0.2% of the number found in a eukaryotic cell.

Ribosomes smaller than those in eukaryotes. They are scattered throughout the cytoplasm, not supported on an endoplasmic reticulum.

Plasmamembrane is a typical phospholipid bilayer.

Food stores are typically lipid globules or glycogen granules.

Flagellum is responsible for motility of many bacteria. It is much simpler than the flagellum of a eukaryotic cell, being composed of a single cylinder of protein subunits (flagellin). The flagellum does not 'beat' but instead rotates about a 'bearing' anchored in the cell wall to produce a corkscrew motion which drives the cell along.

Mesosomes are infoldings of the plasma membrane on which the enzymes associated with respiration are located. A proton gradient generated across these membranes is used to drive the synthesis of ATP.

Lipid structure and function

TRUE LIPIDS are esters of fatty acids and alcohols, formed by condensation reactions. Many of their properties result from

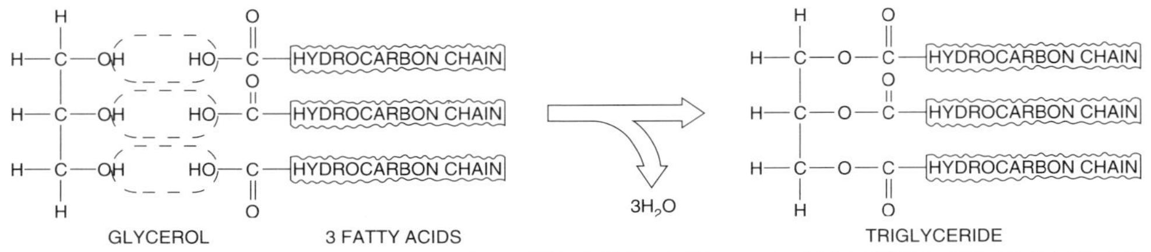

GLYCEROL + 3 FATTY ACIDS → TRIGLYCERIDE + $3H_2O$

HYDROCARBON CHAIN

Since the hydrocarbon chains are long (19 C in arachidonic acid) most of the weight of the triglyceride is fatty acid.

Water-repellent properties: oily secretions of the sebaceous glands help to waterproof the fur and skin. The preen gland of birds produces a secretion which performs a similar function on the feathers.

Cell membranes: phospholipids (phosphatides) are found in all cell membranes. These molecules have a polar 'phosphate-base' group substituted for one of the fatty acids in a triglyceride.

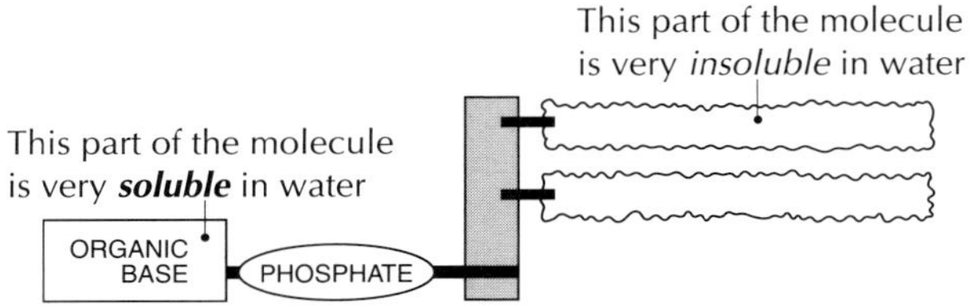

This part of the molecule is very *insoluble* in water

This part of the molecule is very ***soluble*** in water

ORGANIC BASE — PHOSPHATE

Electrical insulation: myelin is secreted by Schwann cells and insulates some neurones in such a way that impulse transmission is made much more rapid.

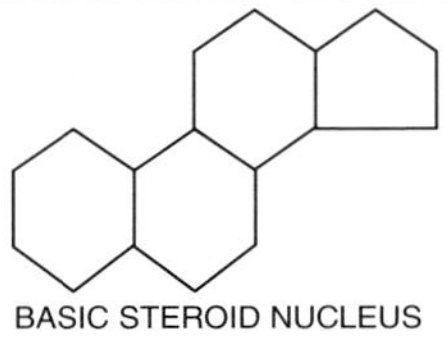

Hormones: an important group of hormones, including cortisone, testosterone and oestrogen, are *steroids*. Steroids are not true esters but have the same solubility properties as them.

BASIC STEROID NUCLEUS

Physical protection: the shock-absorbing ability of subcutaneous fat stores protects delicate organs such as the kidneys from mechanical damage.

Thermal insulation: fats conduct heat very poorly - subcutaneous fat stores help heat retention in endothermic animals. Incompressible blubber is an important insulator in diving mammals.

Attraction: plant scents are derivatives of fatty acids. They are attractive to insects and thus aid pollination.

Waterproofing: the waxy cuticle of insects and plants reduces water losses by evaporation since water cannot cross the insoluble lipid layer. Waxes are esters of higher fatty acids with long chain alcohols (i.e. *not* with glycerol).

Storage: high energy yield per unit mass and insolubility in water make fats and oils ideal energy storage compounds, particularly where dispersal or locomotion requires mass to be kept to a minimum, as in some seeds and fruits.

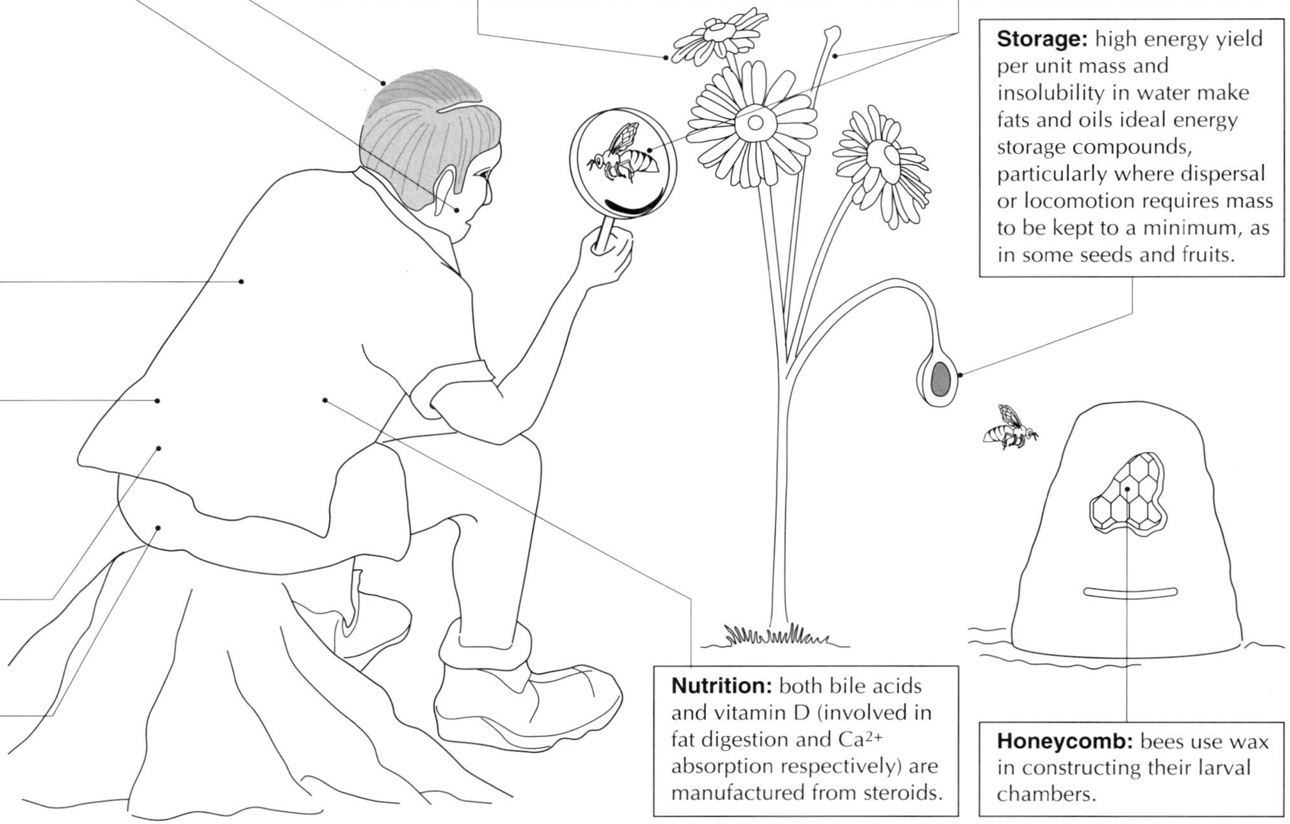

Nutrition: both bile acids and vitamin D (involved in fat digestion and Ca^{2+} absorption respectively) are manufactured from steroids.

Honeycomb: bees use wax in constructing their larval chambers.

Functions of soluble carbohydrates include transport, protection, recognition and energy release.

Sugar derivatives include *sugar alcohols*, e.g. glycerol, *sugar acids*, e.g. ascorbic acid, and *mucopolysaccharides*, which are important components of connective tissues, synovial fluid, cartilage and bone. Heparin (anticoagulant in blood) is derived from mucopolysaccharides and has a protective function.

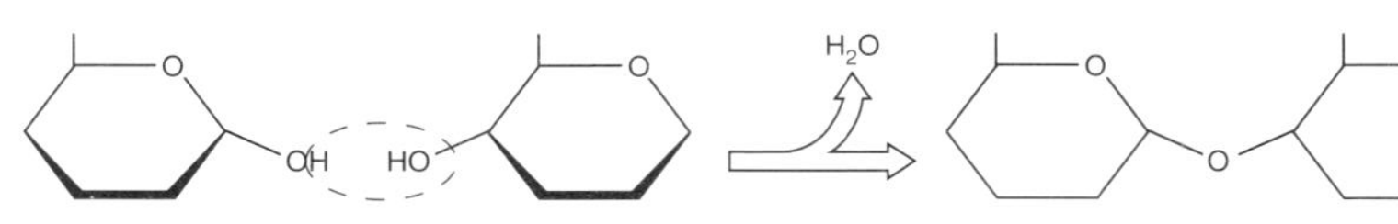

Oligosaccharides are short (often 6-12 units) condensation products which combine with protein (*glycoprotein*) or lipid (*glycolipid*) and form the outer coat (*glycocalyx*) of animal cells. They are important in *cell-cell recognition* and the *immune response*.

INVADER

Glucose is the most common substrate for respiration (energy release). **Fructose** is a constituent of nectar and sweetens fruits to attract animals and aid seed dispersal.

Glucose and fructose are both ***monosaccharides*** (single sugar units) with the typical formula $C_nH_{2n}O_n$. They each have ***six carbon atoms*** and are thus called ***hexoses*** (***pentoses*** have 5 carbon atoms and ***trioses*** have 3).
Glucose and ***fructose*** are isomers of $C_6H_{12}O_6$.

α-GLUCOSE α-FRUCTOSE

In naturally occurring **disaccharides** monosaccharide rings are joined together by *glycosidic bonds.*

H_2O

This most usually occurs between *aldehyde* or *keto group* (i.e. the reducing group) of one monosaccharide and an *hydroxyl group* of another monosaccharide,

e.g. *lactose*

GALACTOSE GLUCOSE

LACTOSE IS A REDUCING DISACCHARIDE

Reducing group of galactose

Hydroxyl group on C_4 of glucose

Reducing group of glucose = carbonyl group (C=O)

(*Maltose* is a reducing disaccharide formed from two molecules of α-glucose.)

or, more rarely, between *reducing groups of adjacent monosaccharides,*

e.g. *sucrose*

GLUCOSE FRUCTOSE

SUCROSE IS A NON-REDUCING DISACCHARIDE

Reducing groups are joined

Ribulose bis phosphate is the *acceptor of CO_2* in the Calvin Cycle.

CO_2 RuBP CALVIN CYCLE 'FIXES' CO_2 TRIOSE SUGAR

Ribose and **deoxyribose** are constituents of *nucleotides*

P ORGANIC BASE

which are the subunits of *nucleic acids* (e.g. DNA).

CHAIN OF NUCLEOTIDES

Other important roles are in the *electron carriers* NAD, FAD and NADP and as the 'energy currency'.

ATP P P P ADENINE

Sucrose (*glucose-fructose*) is the main transport compound in plants. Commonly extracted from sugar cane and sugar beet and used as a sweetener.

TATE & LYLE

Lactose (*glucose-galactose*) is the carbohydrate source for suckling mammals - milk is about 5% lactose.

SEMI SKIM

Maltose (*glucose-glucose*) is a respiratory substrate in germinating seeds.

Polysaccharides are polymers formed by glycosidic bonding of monosaccharide subunits

Starch is a mixture of two polymers of α-glucose: ***amylose*** typically contains about 300 glucose units joined by α 1,4 ***glycosidic bonds***

The bulky $-CH_2OH$ side chains cause the molecule to adopt a helical shape (excellent for packing many subunits into a limited space).

α-glucose molecules

Amylose helix (6 glucose units in each turn)

Amylopectin is a branched chain, containing up to 1500 glucose subunits, in which α 1,4 chains are cross-linked by α 1,6 ***glycosidic bonds.***

α 1,6 GLYCOSIDIC BOND

Because there are so few 'ends' within the starch molecule there are few points to begin hydrolysis by the enzyme ***amylase.*** Starch is therefore an excellent long-term ***storage compound.***

α-GLUCOSE

β-GLUCOSE

Glycogen is an α-glucose polymer, very similar to amylopectin but with very many more cross-links and shorter α 1,4 chains. This is appropriate to animal cells which may need to hydrolyse food reserves more rapidly than plant cells would do.

Cellulose is a polymer of glucose linked by β 1,4 ***glycosidic bonds.*** The β-conformation inverts successive monosaccharide units so that a straight chain polymer is formed.

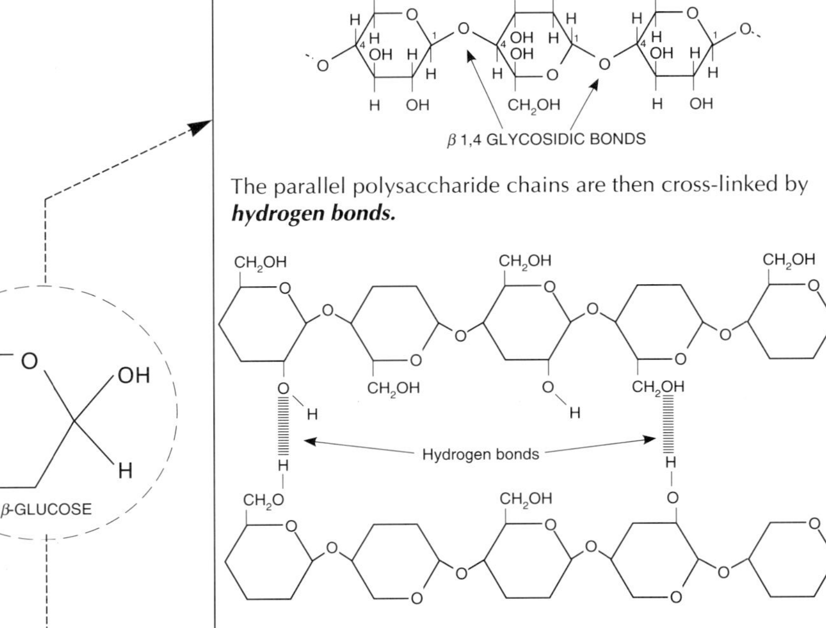

The parallel polysaccharide chains are then cross-linked by ***hydrogen bonds.***

This cross-linking prevents access by water, so that cellulose is very resistant to hydrolysis and is therefore an excellent ***structural molecule*** (cellulose cell walls): ideal in plants which can readily synthesize excess carbohydrate.

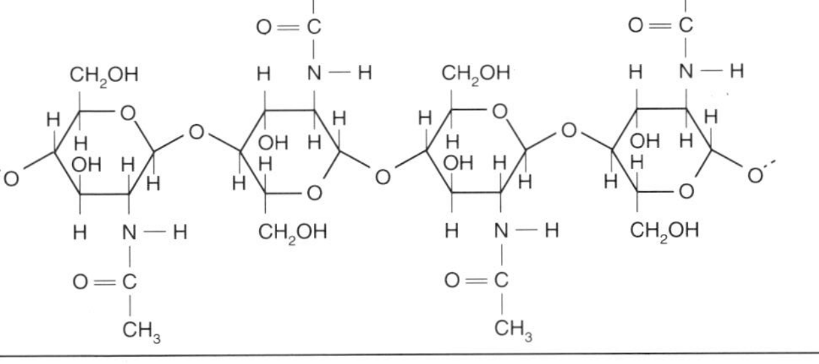

Chitin is another β 1,4 ***polymer*** - the subunits in this case are amino derivatives of β-glucose called ***N-acetylglucosamine.*** It is another ***structural molecule*** abundant in arthropod exoskeletons and in fungal cell walls.

Levels of protein structure

Individual amino acids are joined together by *peptide bonds* by *condensation reactions* catalysed by enzymes. Hydrolysis occurs here during digestion.

CONH is a very important linkage referred to as ***peptide linkage.***

Successive condensations produce a linear chain of amino acids: this *sequence of amino acids* represents the *primary structure of the protein.* This primary structure is maintained by *covalent bonds between adjacent amino acids.*

The polypeptide chains may take on regular arrangements called the *secondary structure* of the protein e.g. the ***α-helix***. This secondary structure is maintained by *hydrogen bonds between the* ***>C=O*** *and >N-H groups of every fourth peptide link.*

An alternative secondary structure - the *β-pleated sheet* - has hydrogen bonds between peptide links of adjacent polypeptide chains.

Sections of α-helix may be folded on themselves: this *supercoiling of the α-helix* represents the *tertiary structure of the protein.* This three-dimensional shape or *conformation* of the protein is maintained by a *series of interactions between -R groups on the polypeptide chain.*

These interactions are very weak so that the conformation of such globular proteins can be easily altered by local physical changes - these alterations are reversible and are essential for the biological function of these molecules.

Several *polypeptide chains* (*tertiary structures*) *may be fitted together* to produce the *quaternary structure of the protein.* The stability of the quaternary structure is maintained by *weak interactions between -R groups of adjacent polypeptide chains* and by *Van der Waal's forces between subunits.*

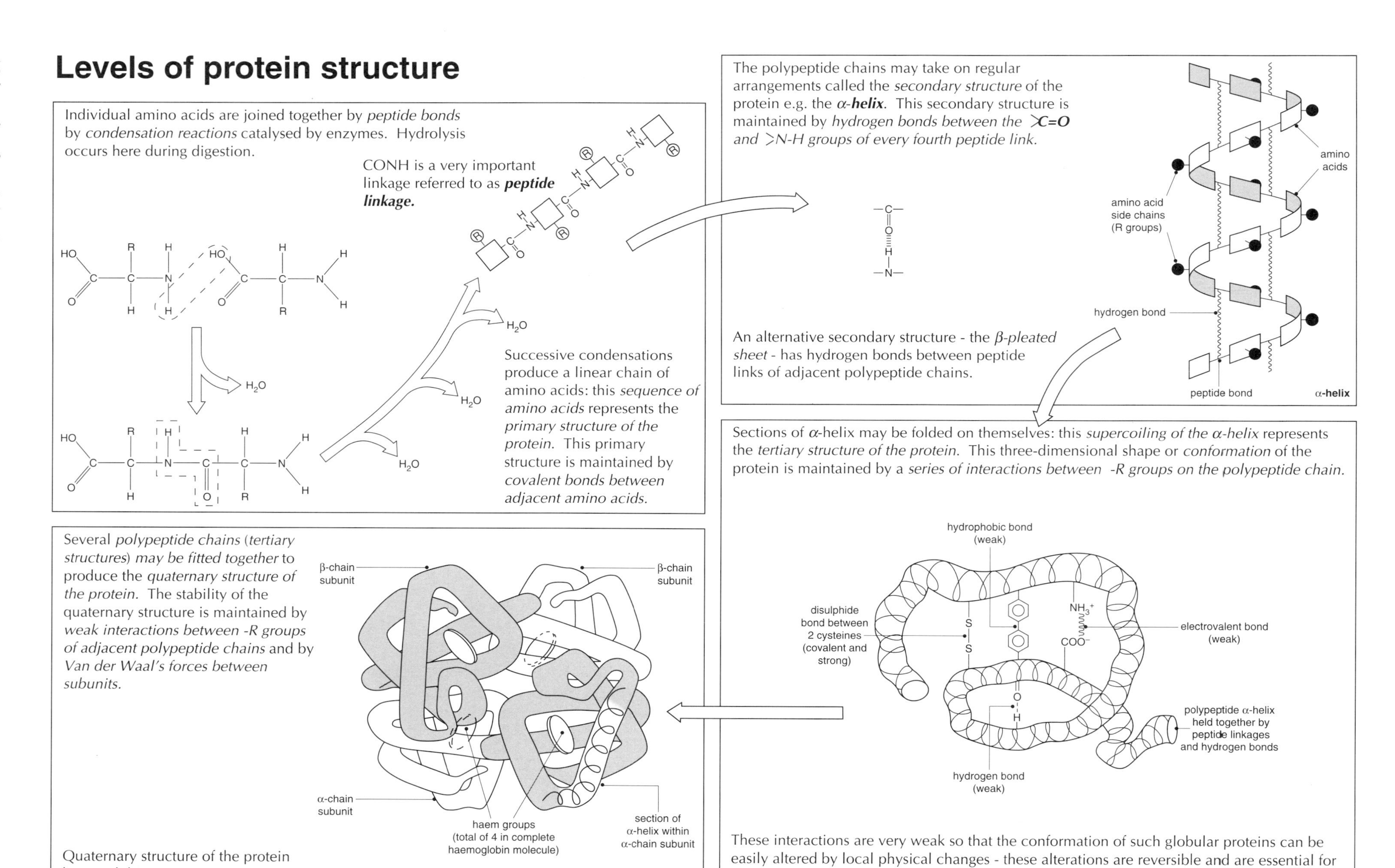

Quaternary structure of the protein haemoglobin.

Functions of proteins

Functions of proteins include transport, catalysis, protection, storage, sensitivity, structure and co-ordination.

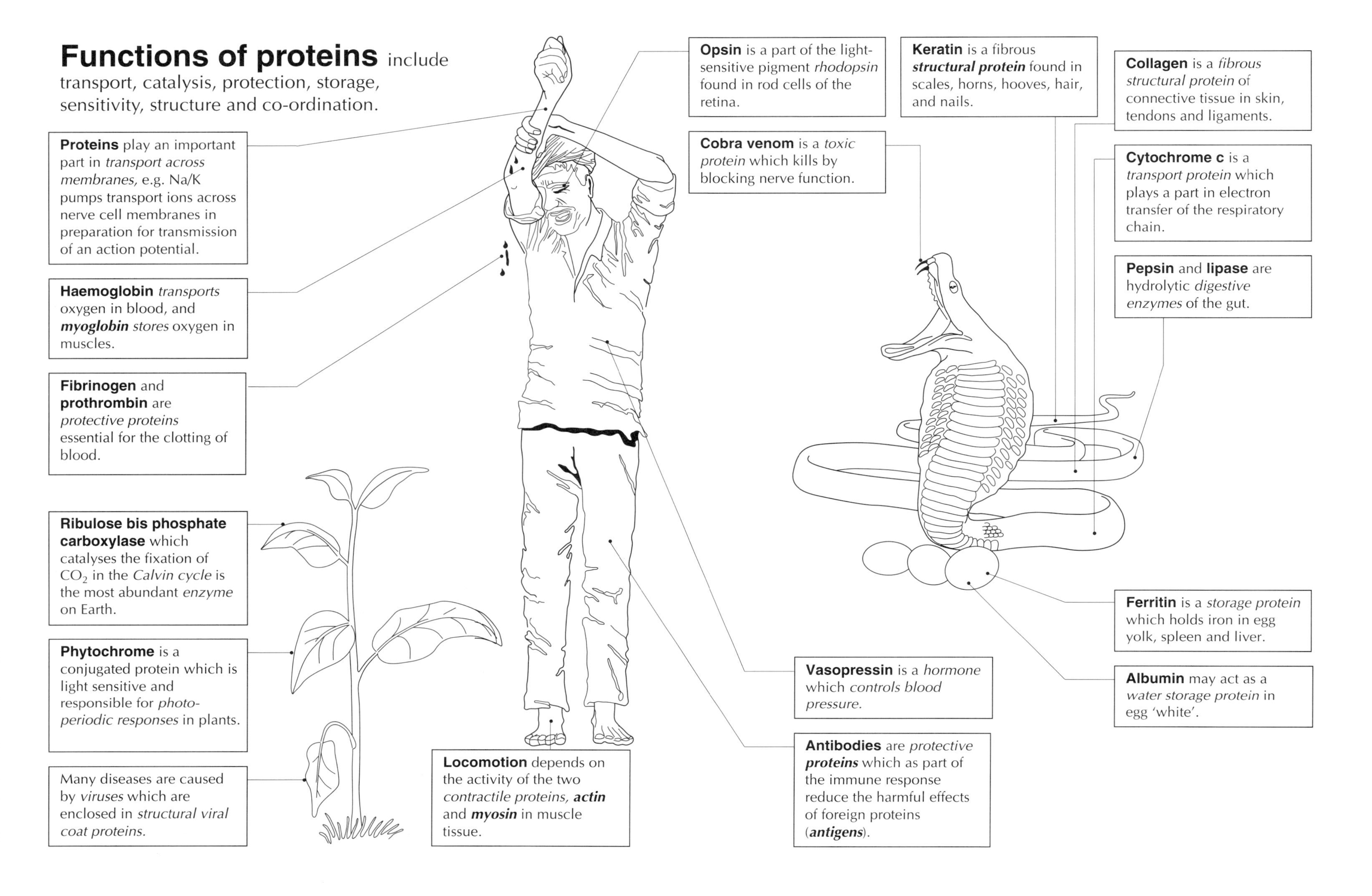

Testing for biochemicals

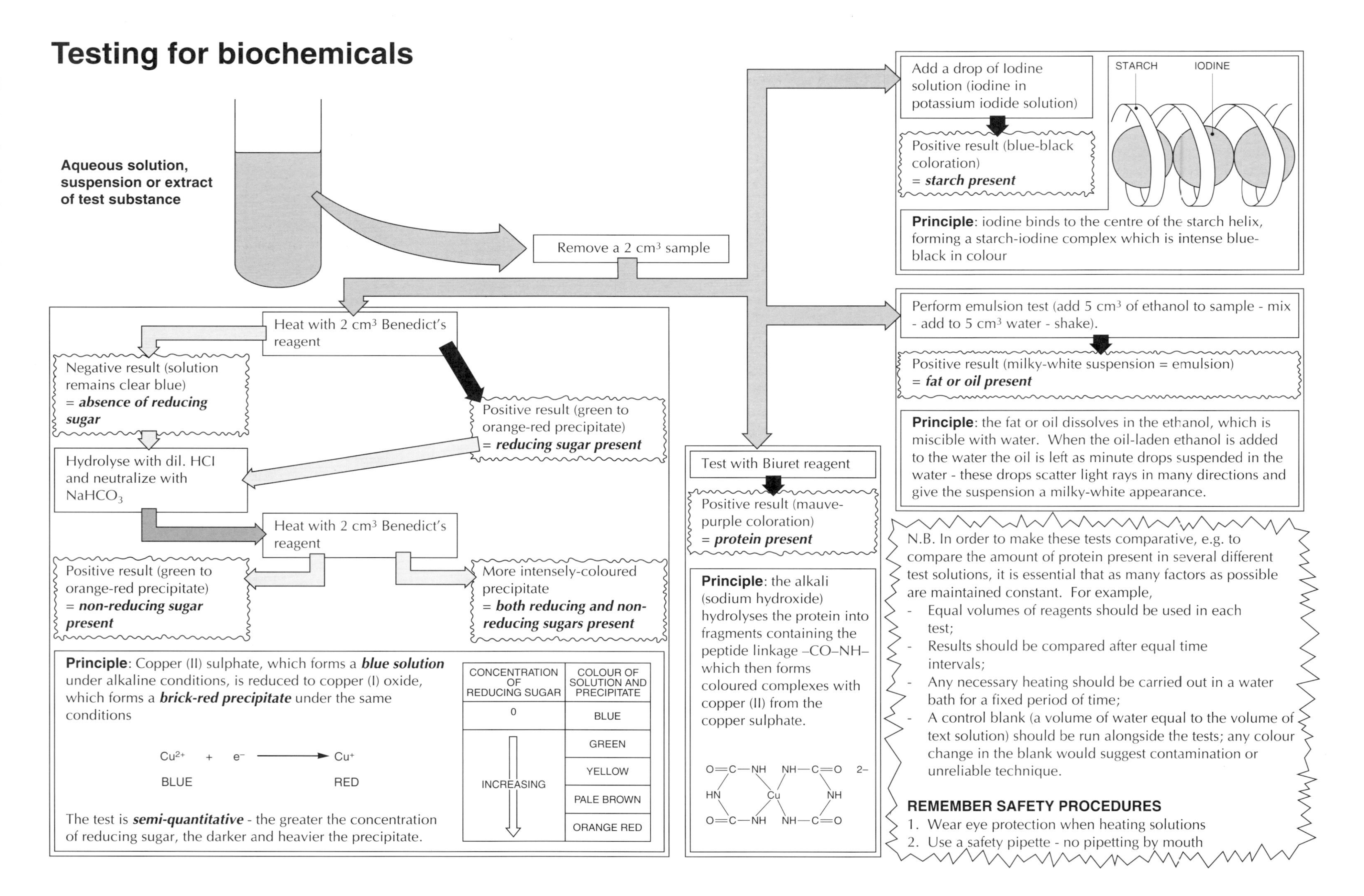

CONCENTRATION OF REDUCING SUGAR	COLOUR OF SOLUTION AND PRECIPITATE
0	BLUE
INCREASING ↓	GREEN
	YELLOW
	PALE BROWN
	ORANGE RED

Catalysis by enzymes

An important step in enzyme catalysis is substrate binding to the active sites.

Enzymes form ***enzyme-substrate complexes*** which reduce the activation energy for reactions which they catalyse.

Consider the reaction: SUBSTRATE (S) $\rightleftharpoons$ PRODUCT (P)

which can be illustrated by a ***reaction profile.***

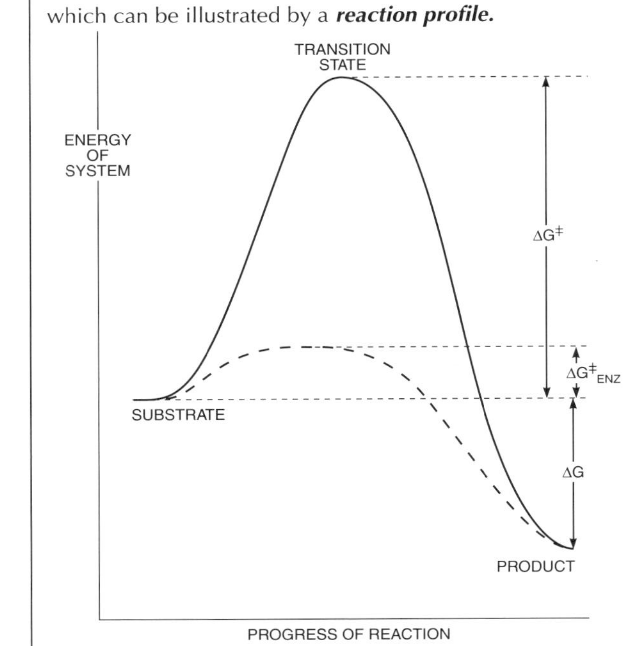

Effect of enzyme on activation energy

For a reaction S⇋P the rate of the forward reaction depends on temperature and activation energy (difference in free energy between substrate and transition state, $\Delta G^{\ddagger}$). The reaction rate is proportional to the number of molecules which have an energy $\geq \Delta G^{\ddagger}$. ***Enzymes act as catalysts by providing alternative reaction pathways in which $\Delta G^{\ddagger}$ is lower than it otherwise would be.*** Heat cannot be used by cells to increase rates of reaction because of possible denaturation.

$$E + S \rightleftharpoons E\text{-}S \rightleftharpoons E + P$$

ENZYME-SUBSTRATE COMPLEX

Stereospecificity: relationship of substrate(s) to active site

Emil Fischer's ***lock and key hypothesis*** suggested that the active site and the substrate were ***exactly complementary***

but more recent work allowed ***Koshland*** to propose the ***induced fit hypothesis*** which suggests that active site and substrate are only fully complementary ***after the substrate is bound.***

This latter process of ***dynamic recognition*** is now the more widely accepted hypothesis.

Cofactors are essential for enzyme activity

Some, such as Zn^{2+} or Mg^{2+}, or porphyrin groups such as the ***haem*** in catalase, may form part of the active site and cannot easily be separated from the enzyme protein: these are commonly called ***prosthetic groups.***

Some, such as NAD (nicotinamide adenine dinucleotide), bind temporarily to the active site and actually take part in the reaction.

e.g. lactate + NAD $\xrightleftharpoons{\text{LACTATE DEHYDROGENASE}}$ pyruvate + $NADH_2$

Such ***coenzymes*** shuttle between one enzyme system and another - most are formed from dietary components called ***vitamins*** (e.g. NAD is formed from niacin, one of the B vitamin complex).

Factors affecting enzyme activity

Factors affecting enzyme activity exert their effects by altering the ease with which an enzyme-substrate complex is formed.

Any factor which alters the conformation (dependent on tertiary structure) of the enzyme will alter the shape of the active site, affect the frequency of enzyme-substrate complex formation and thus influence the rate of the enzyme-catalysed reaction.

EFFECT OF TEMPERATURE

Enzymes have an *optimum temperature* which represents a compromise between *activation* due to increased rate of collision between E and S and *loss of activity* due to denaturation of E molecules and consequent distortion of the active site.

EFFECT OF pH

An enzyme has an *optimum pH* which results from the effects of hydrogen ion concentration on the 3-dimensional shape of the enzyme in the active site region.

Competitive inhibitors compete for the active site with the normal substrate. These inhibitors therefore must have a similar structure to the natural substrate.
The success of the binding of I to the active site depends on the relative concentrations of I and S, and such inhibition is therefore *reversible by an increase in substrate concentration,* e.g. malonate competes with succinate for the active site on the enzyme *succinate dehydrogenase.*

Irreversible inhibition occurs if the enzyme-inhibitor binding is covalent and the distortion of the active site may be permanent, e.g. cyanide (CN^-) binds irreversibly to the active site of the enzyme ***cytochrome oxidase.***

Non-competitive inhibitors reduce enzyme activity by distortion of enzyme conformation caused by binding to some site *other than the active site.* If the binding is non-covalent the inhibition may be *reversible if the inhibitor concentration is diminished.* Many such inhibitors are natural *allosteric regulators* of metabolism, e.g. ATP controls the rate of respiration by inhibition of the enzyme *phosphofructokinase.*

Activators may be necessary to complete the structural relationship between active site and substrate, e.g. chloride ions (Cl^-) are required for activity of the enzyme *salivary amylase.*

There are also *allosteric activators* which enhance enzyme-substrate binding by alteration of enzyme conformation when binding to another (*'allosteric'*) site on the enzyme.

Metabolic pathways help to organize metabolism: each pathway is a series of reactions organized such that the products of one reaction become substrates for the next.

Metabolism – the sum of the chemical reactions within the cell. = **Catabolism** – degradation reactions, some of which release energy and raw materials. + **Anabolism** – synthesis in which large complex molecules are assembled from subunits.

DRIVE

Reactants/precursors are the initial substrates for the metabolic pathway.

Enzymes catalyse the individual steps in a metabolic pathway. These enzymes are highly specific, the first in a metabolic pathway is often subject to allosteric control by an end product.

End products are compounds which the cell can use, store or secrete. These compounds must not be allowed to accumulate, and their concentration commonly regulates the rate of the initial reaction in the pathway which leads to their synthesis by *allosteric control* or *end product inhibition*.

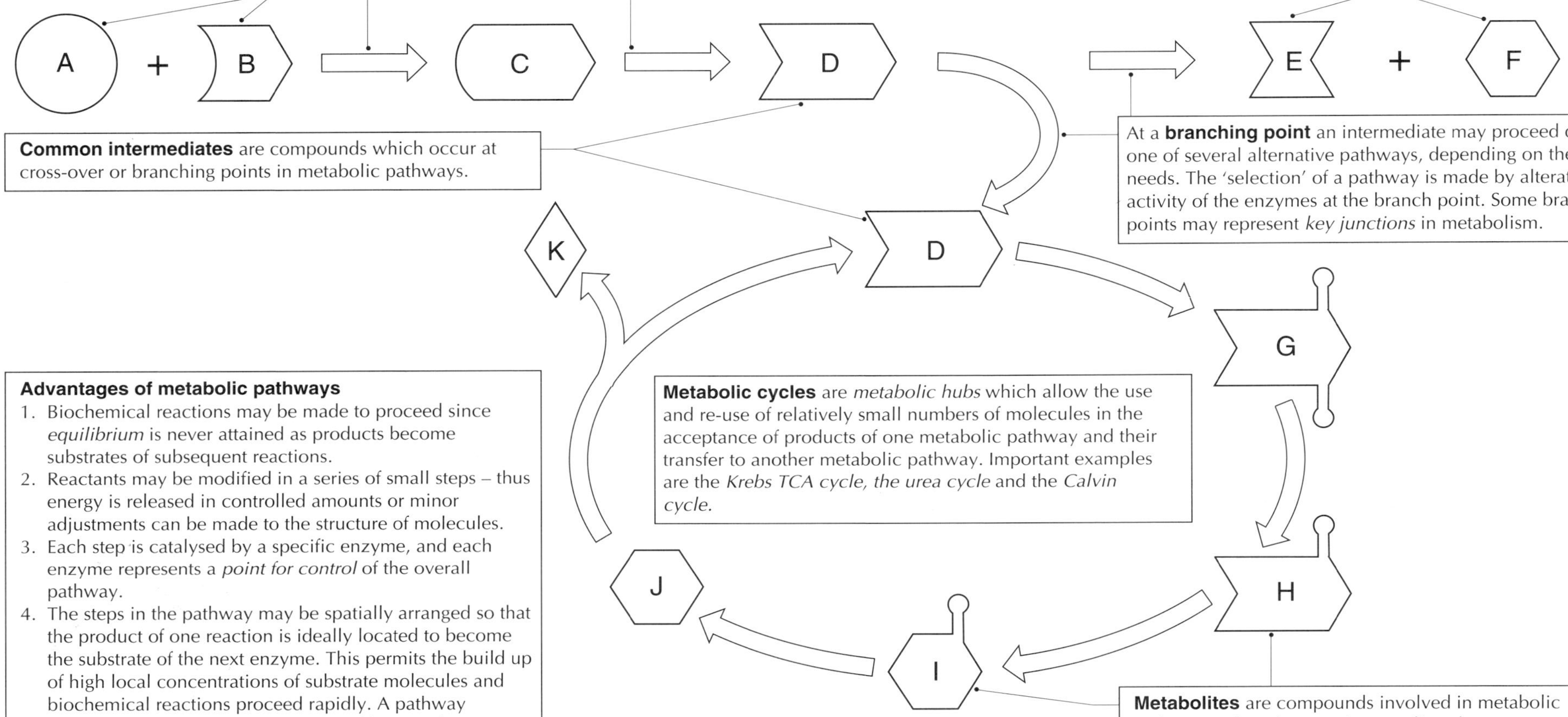

A + B → C → D → E + F

Common intermediates are compounds which occur at cross-over or branching points in metabolic pathways.

At a **branching point** an intermediate may proceed down one of several alternative pathways, depending on the cell's needs. The 'selection' of a pathway is made by alteration of activity of the enzymes at the branch point. Some branching points may represent *key junctions* in metabolism.

D → G → H → I → J → D; J → K

Metabolic cycles are *metabolic hubs* which allow the use and re-use of relatively small numbers of molecules in the acceptance of products of one metabolic pathway and their transfer to another metabolic pathway. Important examples are the *Krebs TCA cycle, the urea cycle* and the *Calvin cycle*.

Metabolites are compounds involved in metabolic pathways; often they are *intermediates* between reactants and end products.

Advantages of metabolic pathways

1. Biochemical reactions may be made to proceed since *equilibrium* is never attained as products become substrates of subsequent reactions.
2. Reactants may be modified in a series of small steps – thus energy is released in controlled amounts or minor adjustments can be made to the structure of molecules.
3. Each step is catalysed by a specific enzyme, and each enzyme represents a *point for control* of the overall pathway.
4. The steps in the pathway may be spatially arranged so that the product of one reaction is ideally located to become the substrate of the next enzyme. This permits the build up of high local concentrations of substrate molecules and biochemical reactions proceed rapidly. A pathway arranged in this way may be catalysed by a *multienzyme complex*.

Commercial applications of enzymes

ANALYSIS

Glucose oxidase
The reaction catalysed by glucose oxidase is

$$\beta\text{-D-glucose} + O_2 \longrightarrow \text{gluconic acid} + H_2O_2$$

The quick and accurate measurement of glucose is of great importance both medically (in sufferers from diabetes, for example) and industrially (in fermentation reactions, for example). A simple quantitative procedure can be devised by coupling the production of hydrogen peroxide to the activity of the enzyme ***peroxidase.***

$$DH_2 + H_2O_2 \xrightarrow{\text{peroxidase}} 2H_2O + D$$

DH_2: chromagen, a hydrogen donor (colourless); D: coloured compound (colour)

Peroxidase can oxidize an organic chromagen (DH_2) to a coloured compound (D) utilizing the hydrogen peroxide – the amount of the coloured compound D produced is a direct measure of the amount of glucose which has reacted. It can be measured quantitatively using a colorimeter or, more subjectively, by comparison with a colour reference card.

CLINISTIX CHART: LOW, NORM, HIGH, DANG

CLINISTIX

This method of glucose analysis is ***highly specific*** and has the enormous advantage over chemical methods in that this specificity allows glucose to be assayed ***in the presence of other sugars***, e.g. in a biological fluid such as blood or urine, without the need for an initial separation.

Both of the enzymes glucose oxidase and peroxidase, and the chromagen DH_2, can be immobilized on a cellulose fibre pad. This forms the basis of the glucose dipsticks ('Clinistix') which were developed to enable diabetics to monitor their own blood or urine glucose levels.

PHARMACEUTICALS ***Papain*** (protein → peptides) is used to remove stains from false teeth.

TEXTILES ***Lipase*** (fats → fatty acids) is used in biological washing powders.

There are many applications of enzyme technology to industry. Enzyme technology has several advantages over 'whole-organism' technology.

a. ***No loss of substrate due to increased biomass.*** For example, when whole yeast is used to ferment sugar to alcohol it always 'wastes' some of the sugar by converting it into cell wall material and protoplasm for its own growth.
b. ***Elimination of wasteful side reactions.*** Whole organisms may convert some of the substrate into irrelevant compounds or even contain enzymes for degrading the desired product into something else.
c. ***Optimum conditions for a particular enzyme may be used.*** These conditions may not be optimal for the whole organism – in some organisms particular enzymes might be working at less than maximum efficiency.
d. ***Purification of the product is easier.*** This is especially true using immobilized enzymes.

MEDICINE

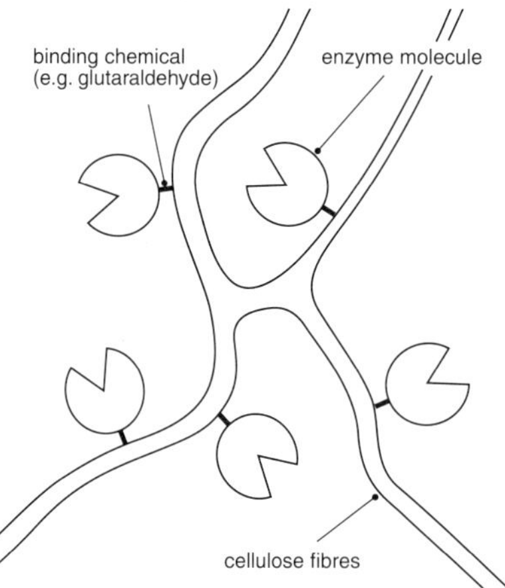

Enzyme immobilization
Immobilization means physically or chemically trapping enzymes or cells onto surfaces or inside fibres. The benefits can be considerable:

a. the same enzyme molecules can be used again and again, since they are not lost;
b. the enzyme does not contaminate the end product;
c. the enzymes may be considerably more stable in immobilized form – for example, glucose isomerase is stable at 65 °C when immobilized.

STERILIZED SKIMMED MILK

IMMOBILIZED YEAST LACTASE hydrolyses lactose → glucose and galactose

LACTOSE–FREE MILK

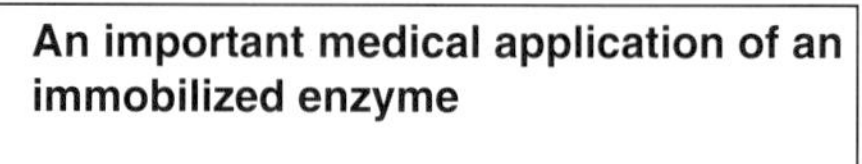

An important medical application of an immobilized enzyme

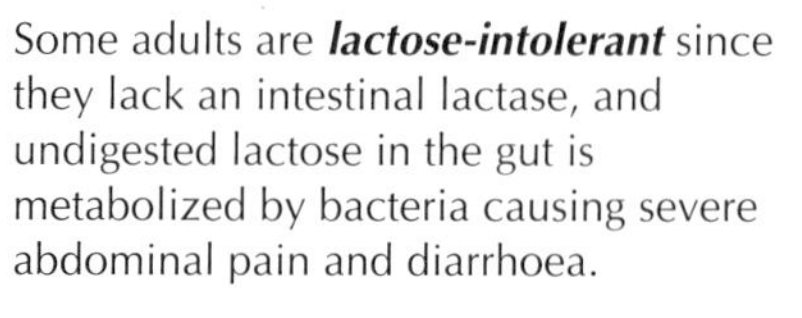

Some adults are ***lactose-intolerant*** since they lack an intestinal lactase, and undigested lactose in the gut is metabolized by bacteria causing severe abdominal pain and diarrhoea.

Milk is an important dietary component and can be made ***lactose-free*** by passage down a column packed with ***yeast lactase*** immobilized on fibres of cellulose acetate.

Glycolysis generates ATP, reduced electron carriers and pyruvate.

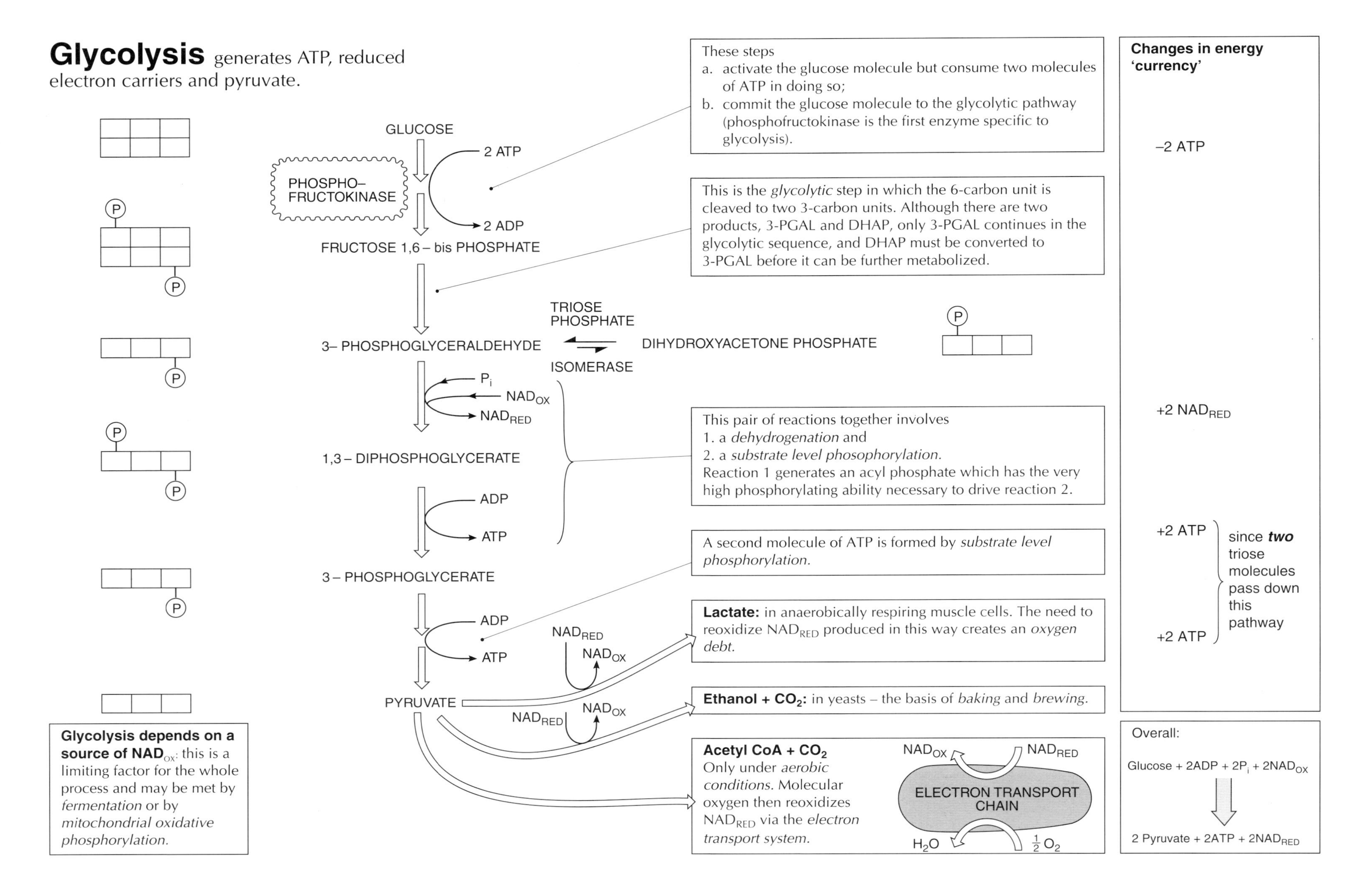

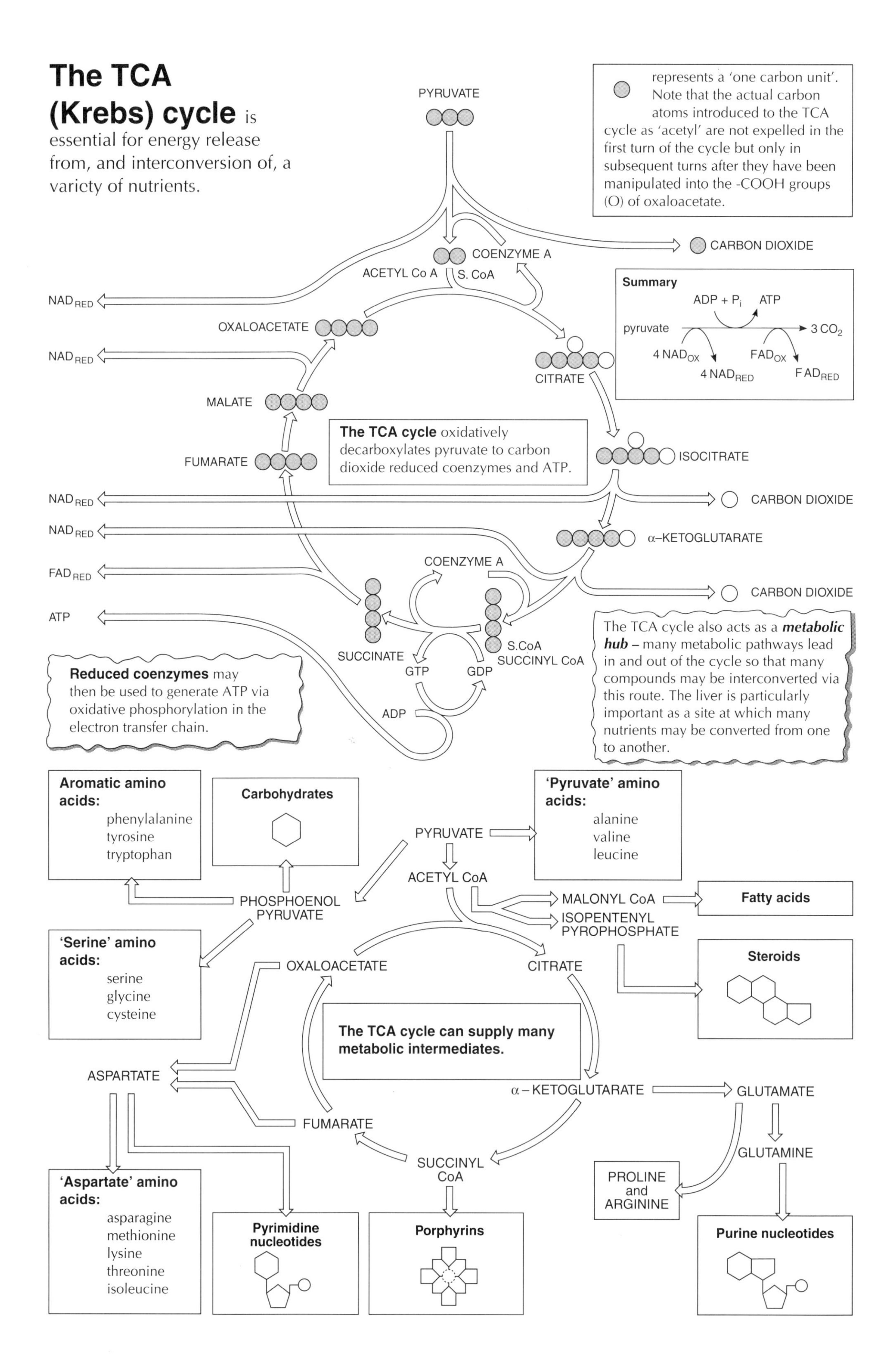
The TCA (Krebs) cycle is essential for energy release from, and interconversion of, a variety of nutrients.
PYRUVATE
represents a 'one carbon unit'. Note that the actual carbon atoms introduced to the TCA cycle as 'acetyl' are not expelled in the first turn of the cycle but only in subsequent turns after they have been manipulated into the -COOH groups (O) of oxaloacetate.
CARBON DIOXIDE
COENZYME A
ACETYL Co A
S. CoA
NAD RED
OXALOACETATE
NAD RED
MALATE
FUMARATE
CITRATE
ISOCITRATE
Summary
ADP + Pi
ATP
pyruvate
3 CO2
4 NADOX
4 NADRED
FADOX
FADRED
The TCA cycle oxidatively decarboxylates pyruvate to carbon dioxide reduced coenzymes and ATP.
NAD RED
CARBON DIOXIDE
NAD RED
α–KETOGLUTARATE
COENZYME A
FAD RED
CARBON DIOXIDE
ATP
SUCCINATE
S.CoA
SUCCINYL CoA
GTP
GDP
ADP
The TCA cycle also acts as a metabolic hub – many metabolic pathways lead in and out of the cycle so that many compounds may be interconverted via this route. The liver is particularly important as a site at which many nutrients may be converted from one to another.
Reduced coenzymes may then be used to generate ATP via oxidative phosphorylation in the electron transfer chain.
Aromatic amino acids:
phenylalanine
tyrosine
tryptophan
Carbohydrates
PYRUVATE
'Pyruvate' amino acids:
alanine
valine
leucine
ACETYL CoA
PHOSPHOENOL PYRUVATE
MALONYL CoA
Fatty acids
ISOPENTENYL PYROPHOSPHATE
'Serine' amino acids:
serine
glycine
cysteine
OXALOACETATE
CITRATE
Steroids
The TCA cycle can supply many metabolic intermediates.
ASPARTATE
α – KETOGLUTARATE
GLUTAMATE
FUMARATE
GLUTAMINE
SUCCINYL CoA
PROLINE and ARGININE
'Aspartate' amino acids:
asparagine
methionine
lysine
threonine
isoleucine
Pyrimidine nucleotides
Porphyrins
Purine nucleotides

Cellular respiration occurs in a series of localized stages.

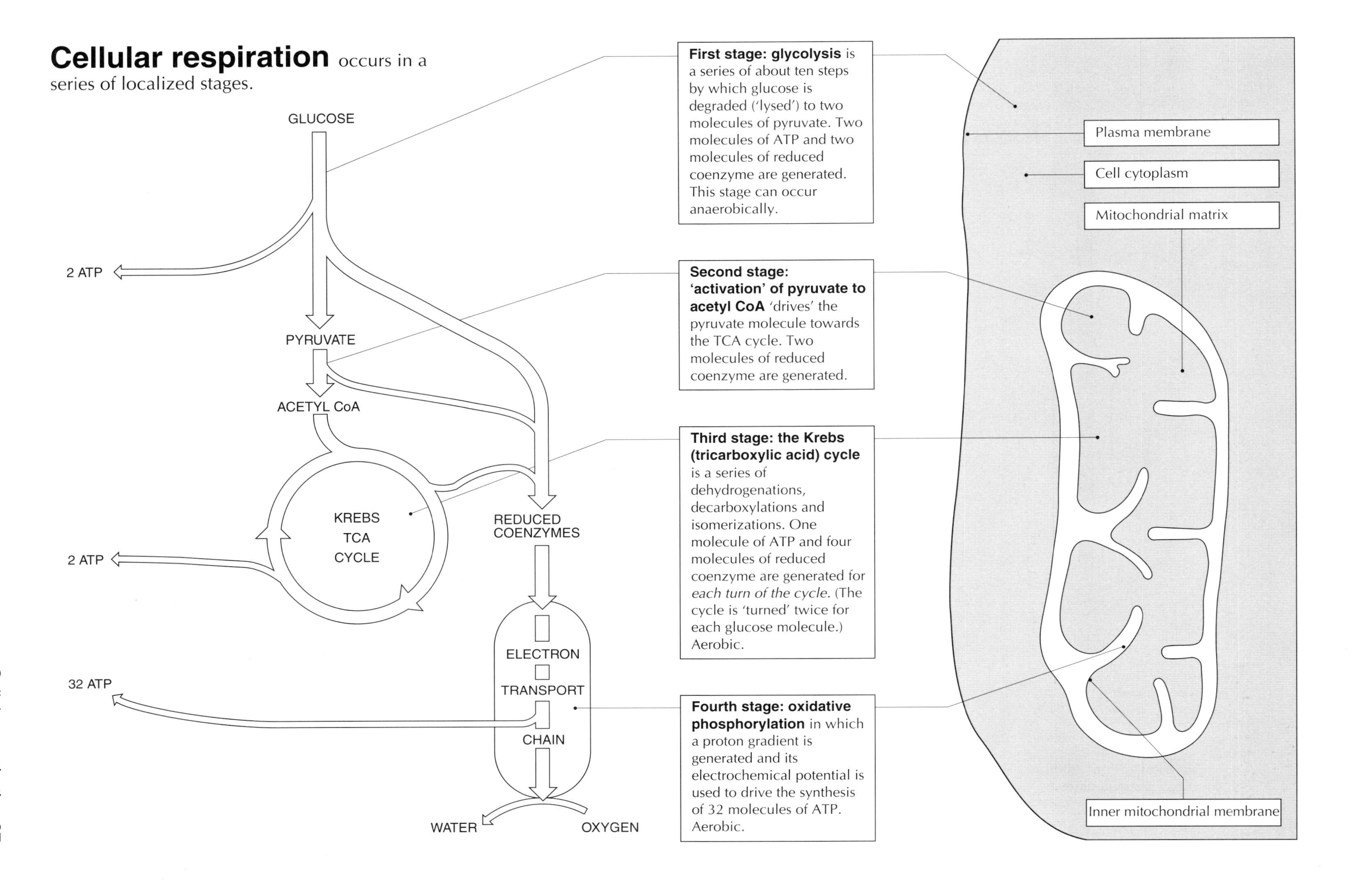

First stage: glycolysis is a series of about ten steps by which glucose is degraded ('lysed') to two molecules of pyruvate. Two molecules of ATP and two molecules of reduced coenzyme are generated. This stage can occur anaerobically.

Second stage: 'activation' of pyruvate to acetyl CoA 'drives' the pyruvate molecule towards the TCA cycle. Two molecules of reduced coenzyme are generated.

Third stage: the Krebs (tricarboxylic acid) cycle is a series of dehydrogenations, decarboxylations and isomerizations. One molecule of ATP and four molecules of reduced coenzyme are generated for *each turn of the cycle.* (The cycle is 'turned' twice for each glucose molecule.) Aerobic.

Fourth stage: oxidative phosphorylation in which a proton gradient is generated and its electrochemical potential is used to drive the synthesis of 32 molecules of ATP. Aerobic.

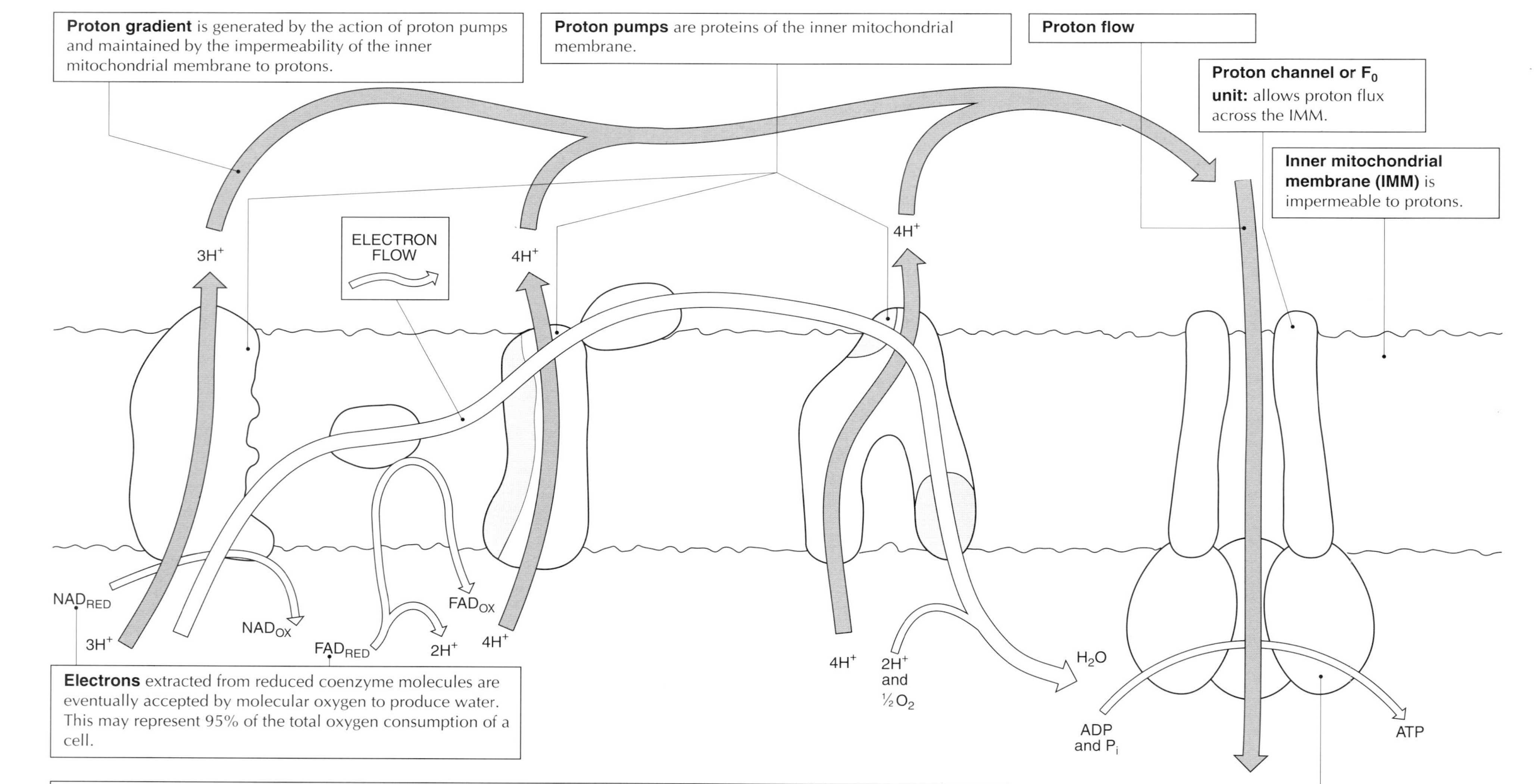

Mitchell's chemiosmotic theory suggests that:

1. there is an *exergonic series* of redox reactions between members of the electron transport chain on the inner mitochondrial membrane;
2. these exergonic redox reactions can drive three proton pumps which transport H^+ from the mitochondrial matrix to the inter-membrane space, thus generating a *proton gradient* across the IMM;
3. dissipation of proton gradient can be coupled to the phosphorylation of ADP to ATP.

ATP synthetase or F_1 unit couples the flow of protons down the proton concentration gradient to the phosphorylation of ADP. It has been suggested that the proton flux through the enzyme alters the nucleotide-binding properties of the active site, promoting ATP synthesis.

ATP: the energy currency of the cell

ATP hydrolysis is favoured

$$ATP^{4-} + H_2O \rightleftharpoons ADP^{3-} + P^{2-} + H^+ + 30.5 \text{ kJ mol}^{-1}$$

i.e. ATP has a strong tendency to transfer its terminal phosphoryl group, a reaction associated with the release of 30.5 kJ mol^{-1} of ATP, because

1. the repulsion between the four negative charges in ATP^{4-} is reduced when ATP is hydrolysed because two negative charges are removed with phosphate.
2. the H^+ ion which is released when ATP is hydrolysed reacts with OH^- ions to form water – this is a highly favoured reaction.
3. the charge distribution on ADP + P is more stable than that on ATP.

This part of the molecule acts like a 'handle' – it has a shape which can be recognized by highly specific enzymes.

This part of the molecule contains anhydride bonds (O-P) which can be hydrolysed in reactions which are ***exergonic*** (energy-yielding) and can be coupled to ***endergonic*** (energy-demanding) reactions.

NH_2

ADENINE

RIBOSE

OH OH

CH_2

ADENOSINE

ADENOSINE TRIPHOSPHATE

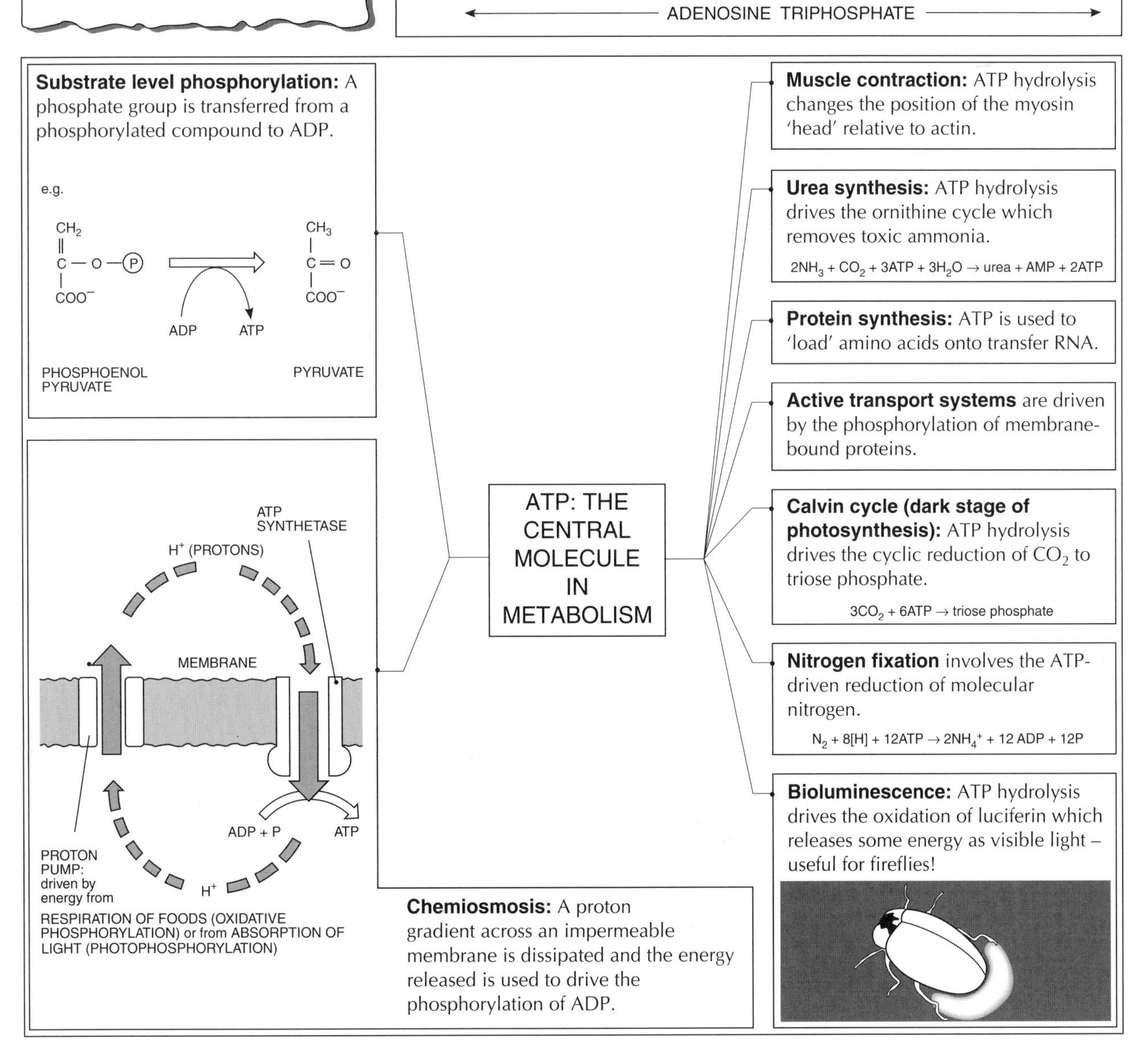

Substrate level phosphorylation: A phosphate group is transferred from a phosphorylated compound to ADP.

Chemiosmosis: A proton gradient across an impermeable membrane is dissipated and the energy released is used to drive the phosphorylation of ADP.

Muscle contraction: ATP hydrolysis changes the position of the myosin 'head' relative to actin.

Urea synthesis: ATP hydrolysis drives the ornithine cycle which removes toxic ammonia.

$2NH_3 + CO_2 + 3ATP + 3H_2O \rightarrow \text{urea} + AMP + 2ATP$

Protein synthesis: ATP is used to 'load' amino acids onto transfer RNA.

Active transport systems are driven by the phosphorylation of membrane-bound proteins.

Calvin cycle (dark stage of photosynthesis): ATP hydrolysis drives the cyclic reduction of CO_2 to triose phosphate.

$3CO_2 + 6ATP \rightarrow \text{triose phosphate}$

Nitrogen fixation involves the ATP-driven reduction of molecular nitrogen.

$N_2 + 8[H] + 12ATP \rightarrow 2NH_4^+ + 12\ ADP + 12P$

Bioluminescence: ATP hydrolysis drives the oxidation of luciferin which releases some energy as visible light – useful for fireflies!

Nucleic acids I: DNA
Deoxyribnucleic acid

The Watson–Crick model for DNA suggests that the molecule is a double helix of two complementary, anti-parallel polynucleotide chains

Nucleotides are the subunits of nucleic acids, including DNA. Each of these subunits is made up of:

AN ORGANIC (NITROGENOUS) BASE
+
A PENTOSE SUGAR
+
A PHOSPHATE GROUP (PHOSPHORIC ACID)

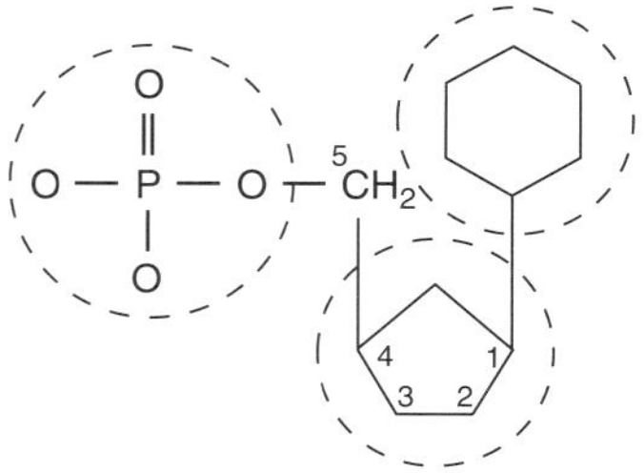

Note that the phosphate group is bonded to the C_5 atom of the pentose sugar.

There are *four different nucleotides* in a DNA molecule; they differ only in the organic (nitrogen) base present.

There are two different *pyrimidine (single ring) bases*, called *cytosine (C)* and *thymine (T)*

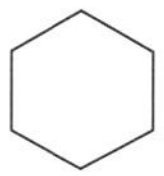

and

two different *purine (double ring) bases* called *adenine (A)* and *guanine (G).*

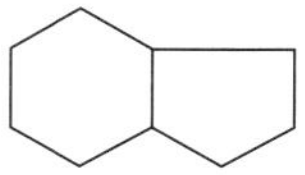

The different dimensions of the purine and pyrimidine bases is extremely important in the formation of the double-stranded DNA molecule.

Nucleotides are linked to form a *polynucleotide* by the formation of 3′ 5′ *phosphodiester links* in which a phosphate group forms a bridge between the C_3 of one sugar molecule and the C_5 of the next sugar molecule.

There are *ten base pairs* for each pitch of the double helix.

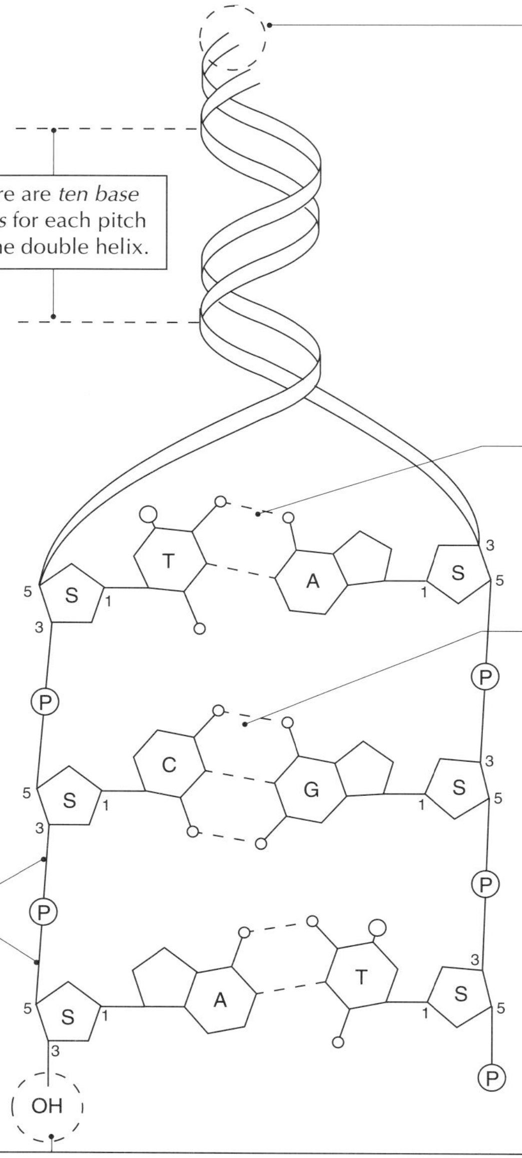

This is the 5′ *end of the chain* since the terminal phosphate group is only bonded to the C_5 of the sugar molecule.

Base pairing in DNA was proposed to explain how two polynucleotide chains could be held together by hydrogen bonds. To accommodate the measured dimensions of the molecule each base pair comprises *one purine - one pyrimidine.*

The double helix is most stable, that is the greatest number of hydrogen bonds is formed, when the base pairs

A :::::::: T (two hydrogen bonds)

and G :::::::: C (three hydrogen bonds)

are formed. These are *complementary base pairs.*

Note that in order to form and maintain this number of hydrogen bonds the nucleotides are inverted with respect to one another so that the phosphate groups (here shown as Ⓟ) *face in opposite directions.*

The chains are *anti-parallel,* that is one chain runs from 5′→3′ whilst the other runs from 3′→5′.

The chains are *complementary:* because of base pairing, the base sequence on one of the chains automatically dictates the base sequence on the other.

This is the 3′ *end of the chain* since the C_3 atom of the sugar molecule of the final nucleotide has a ‘free’ -OH group which is not part of a phosphodiester link.

Nucleic acids II: RNA ***Ribonucleic acid*** has a number of functions in protein synthesis.

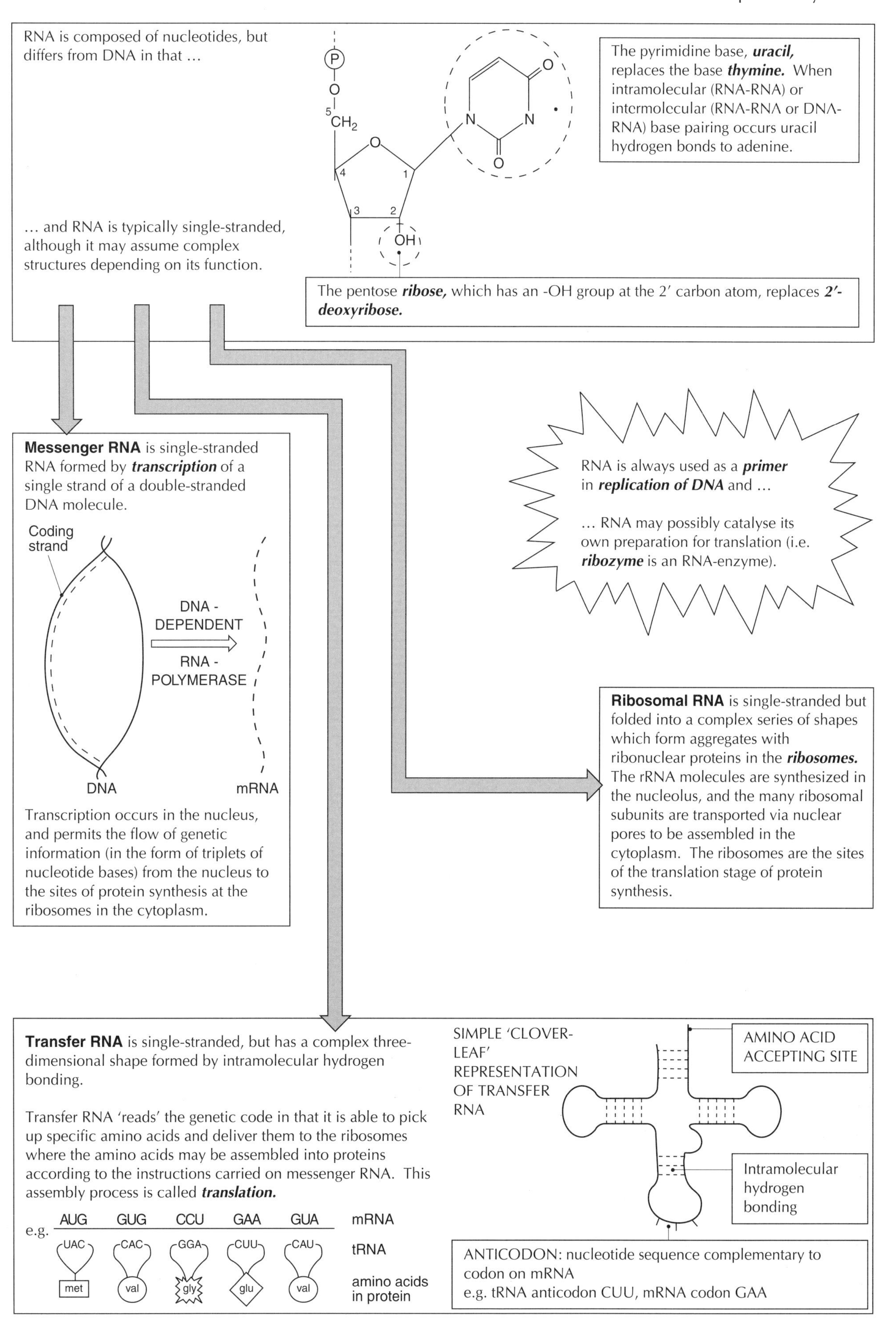

Leaf structure is adapted for photosynthesis and for gas exchange.

Phototropism is a growth response which allows shoots to grow towards the light to allow optimum illumination of the leaves.

Large leaf surface area is held perpendicular to the light source (and kept there by a 'tracking' system).

Leaves are thin so that there are few cell layers to absorb light before it is received by the photosynthetic cells.

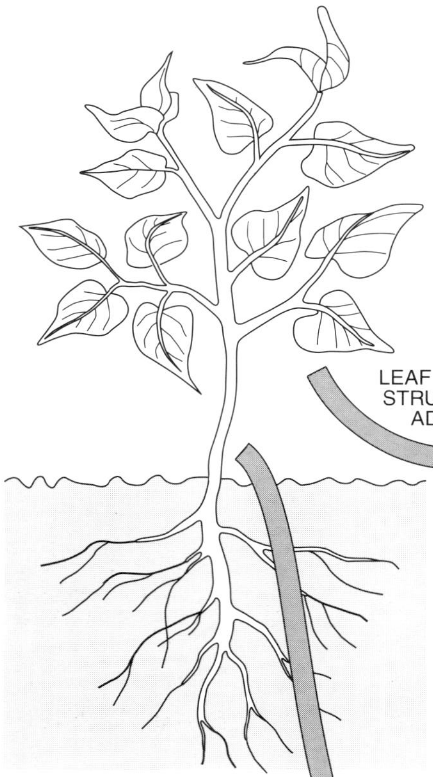

Leaf mosaic is the arrangement of leaves in a pattern which minimizes overlapping/shading but maximizes leaf exposure to light.

Shoot system holds leaves in optimum position for illumination and CO_2 uptake.

Etiolation causes rapid elongation of internodes kept in darkness to extend shoot.

LEAF INTERNAL STRUCTURE IS ADAPTED

GAS EXCHANGE IN STEMS IS MADE POSSIBLE BY LENTICELS

Cuticle: is composed of a waxy compound called ***cutin*** and is secreted by the epidermis. It is the cuticle which reduces water loss by evaporation, not the epidermis itself. May be considerably thickened in xerophytes.

Upper epidermis: is one or two cells thick – protects against ***water loss*** (either by cuticle – see left – or by epidermal hairs which trap moisture and reflect light) and against ***invasion by pathogens*** so that the moist inner leaf is quite sterile. Transparent to visible light.

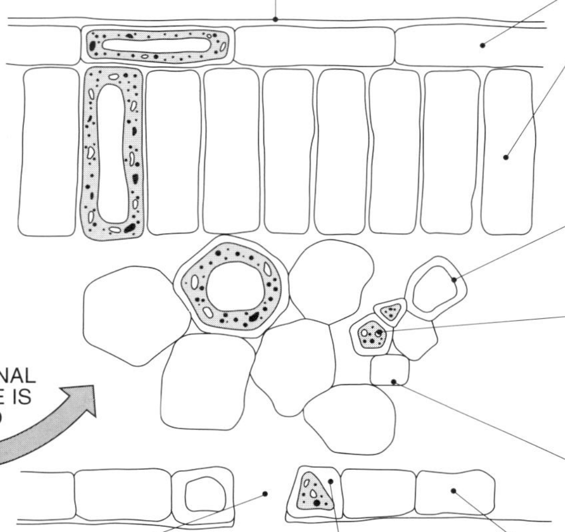

Palisade mesophyll: is the major site of photosynthesis – cells packed vertically with many chloroplasts which may move by cytoplasmic streaming to optimum position within the cell for light absorption and subsequent photosynthesis.

Xylem vessel: transports water and mineral salts to the leaves. Heavily lignified cell walls help to maintain extension of the leaf blade.

Phloem sieve tube: removes products of photosynthesis (principally sucrose) and may import other organic solutes such as amino acids/amides and help to redistribute ions such as phosphate.

Spongy mesophyll: irregularly shaped cells which fit together loosely to leave large air spaces which permit diffusion of gases through leaves. There is much evaporative water loss from the surface of these cells.

Stomatal pore can be opened (to allow diffusion of O_2 and CO_2 down concentration gradients) or closed (to limit water losses by evaporation to a drier atmosphere).

Lower epidermis: similar protective functions to upper epidermis. Cuticle usually thinner (less light → lower temperature → lower evaporation rate).

Guard cell: has chloroplasts and membrane proteins to permit pumping of K^+ ions to drive osmotic movement of water. Uneven thickening of cellulose cell wall permits opening/closure of stomatal pore as turgidity changes.

Lenticel

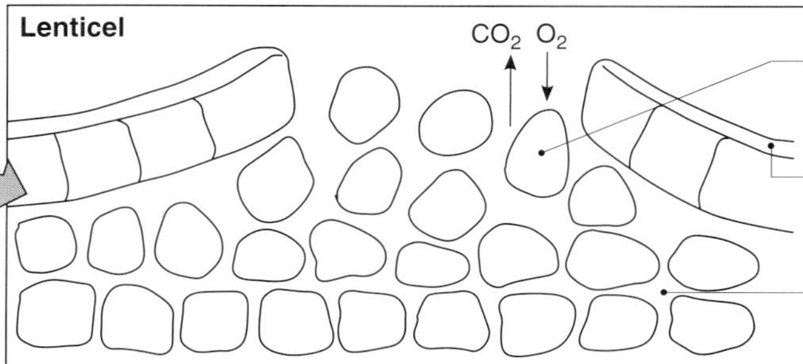

CO_2 O_2

Loosely packed cork cells with moist covering permit exchange of gases between tissues and environment.

Epidermis with waterproof cuticle prevents gas exchange between stem and environment.

Intercellular air spaces permit free movement of gases through plant body.

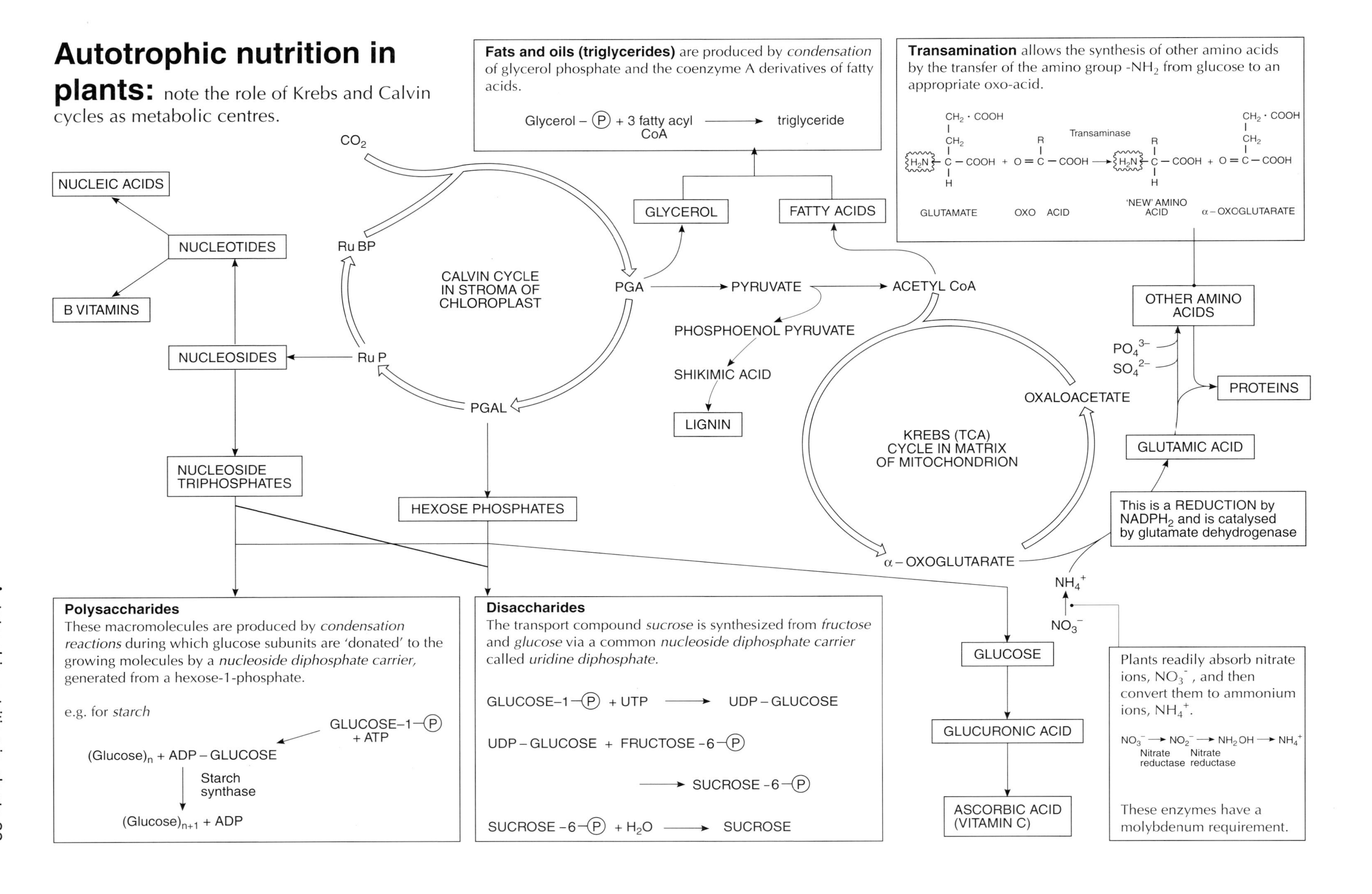

Autotrophic nutrition in plants: note the role of Krebs and Calvin cycles as metabolic centres.
Fats and oils (triglycerides) are produced by condensation of glycerol phosphate and the coenzyme A derivatives of fatty acids.
Glycerol – (P) + 3 fatty acyl CoA → triglyceride
Transamination allows the synthesis of other amino acids by the transfer of the amino group -NH_2 from glucose to an appropriate oxo-acid.
$CH_2 \cdot COOH$
CH_2
H_2N – C – COOH + O = C – COOH → H_2N – C – COOH + O = C – COOH
Transaminase
GLUTAMATE
OXO ACID
'NEW' AMINO ACID
α – OXOGLUTARATE
CO_2
NUCLEIC ACIDS
NUCLEOTIDES
B VITAMINS
NUCLEOSIDES
Ru BP
Ru P
CALVIN CYCLE IN STROMA OF CHLOROPLAST
PGA
PGAL
GLYCEROL
FATTY ACIDS
PYRUVATE
ACETYL CoA
PHOSPHOENOL PYRUVATE
SHIKIMIC ACID
LIGNIN
OTHER AMINO ACIDS
PO_4^{3-}
SO_4^{2-}
PROTEINS
OXALOACETATE
KREBS (TCA) CYCLE IN MATRIX OF MITOCHONDRION
GLUTAMIC ACID
NUCLEOSIDE TRIPHOSPHATES
HEXOSE PHOSPHATES
This is a REDUCTION by $NADPH_2$ and is catalysed by glutamate dehydrogenase
α – OXOGLUTARATE
NH_4^+
NO_3^-
Polysaccharides
These macromolecules are produced by condensation reactions during which glucose subunits are 'donated' to the growing molecules by a nucleoside diphosphate carrier, generated from a hexose-1-phosphate.
e.g. for starch
GLUCOSE–1–(P) + ATP
$(Glucose)_n$ + ADP – GLUCOSE
Starch synthase
$(Glucose)_{n+1}$ + ADP
Disaccharides
The transport compound sucrose is synthesized from fructose and glucose via a common nucleoside diphosphate carrier called uridine diphosphate.
GLUCOSE–1–(P) + UTP → UDP – GLUCOSE
UDP – GLUCOSE + FRUCTOSE -6–(P)
→ SUCROSE -6–(P)
SUCROSE -6–(P) + H_2O → SUCROSE
GLUCOSE
GLUCURONIC ACID
ASCORBIC ACID (VITAMIN C)
Plants readily absorb nitrate ions, NO_3^-, and then convert them to ammonium ions, NH_4^+.
NO_3^- → NO_2^- → NH_2OH → NH_4^+
Nitrate reductase
Nitrate reductase
These enzymes have a molybdenum requirement.

Law of limiting factors

Blackman stated: 'when a process is affected by more than one factor its rate is limited by the factor which is nearest its minimum value: it is that *limiting factor* which directly affects a process if its magnitude is changed.'

Photosynthesis is a multi-stage process – for example the Calvin cycle is dependent on the supply of ATP and reducing power from the light reactions – and the principle of limiting factors can be applied.

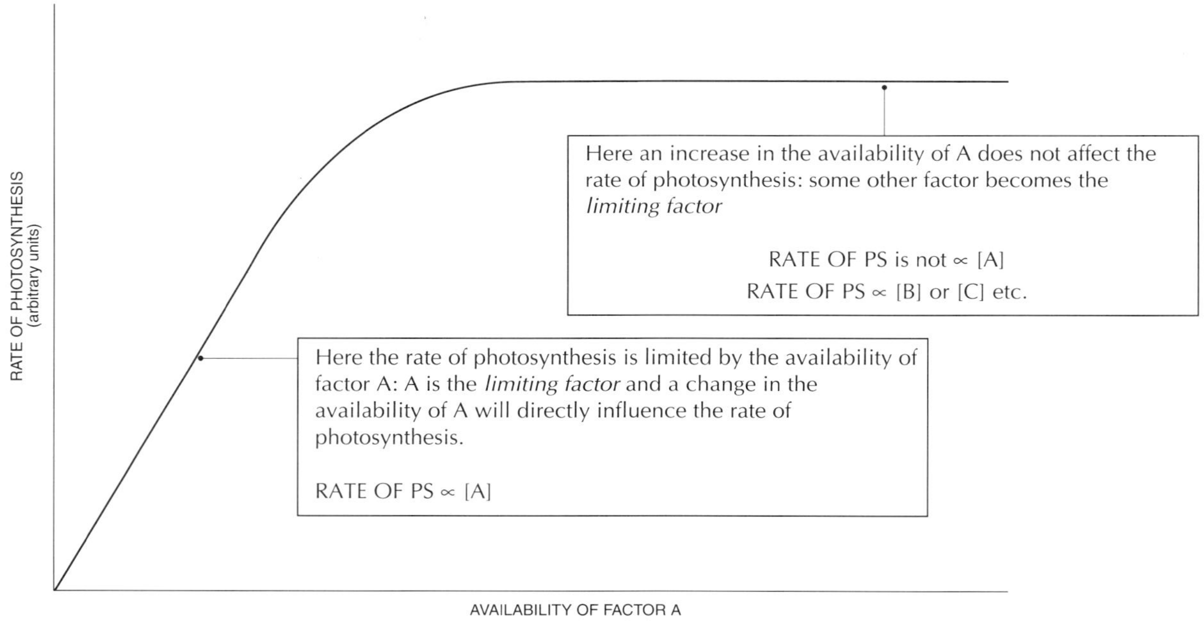

RATE OF PHOTOSYNTHESIS (arbitrary units)

AVAILABILITY OF FACTOR A

Here the rate of photosynthesis is limited by the availability of factor A: A is the *limiting factor* and a change in the availability of A will directly influence the rate of photosynthesis.

RATE OF PS $\propto$ [A]

Here an increase in the availability of A does not affect the rate of photosynthesis: some other factor becomes the *limiting factor*

RATE OF PS is not $\propto$ [A]
RATE OF PS $\propto$ [B] or [C] etc.

The rate of a multi-stage process may be subject to different limiting factors at different times. Photosynthesis may be limited by *temperature* during the early part of a summer's day, by *light intensity* during cloudy or overcast conditions or by *carbon dioxide concentration* at other times. The principal limiting factor in Britain during the summer is *carbon dioxide concentration:* the atmospheric [CO_2] is typically only 0.04%. Increased CO_2 emissions from combustion of fossil fuels may stimulate photosynthesis.

The *mechanism of photosynthesis* is made clearer by studies of limiting factors – the fact that *light* is a limiting factor indicates a *light-dependent stage*, the effect of *temperature* suggests that there are *enzyme-catalysed* reactions, the *interaction of [CO_2]* and *temperature* suggests an enzyme catalysed *fixation of carbon dioxide*. The existence of more than one limiting factor suggests that *photosynthesis is a multi-stage process.*

The *limiting factors* which affect *photosynthesis* are:

Light intensity: light energy is necessary to generate ATP and $NADPH_2$ during the light dependent stages of photosynthesis.

Carbon dioxide concentration: CO_2 is 'fixed' by reaction with ribulose bisphosphate in the initial reaction of the Calvin cycle.

Temperature: the enzymes catalysing the reactions of the Calvin cycle and some of the light-dependent stages are affected by temperature.

Water availability and **chlorophyll concentration** are not normally limiting factors in photosynthesis.

The study of limiting factors has *commercial and horticultural applications*. Since [CO_2] is a limiting factor, crop production in greenhouses is readily stimulated by raising local carbon dioxide concentrations (from gas cylinders or by burning fossil fuels). It is also clear to horticulturalists that expensive increases in energy consumption for lighting and heating are not economically justified if neither of these is the limiting factor applying under any particular set of conditions.

Light reaction: non-cyclic photophosphorylation

Light energy excites electrons, resulting in the splitting of water and the synthesis of ATP and $NADPH_2$.

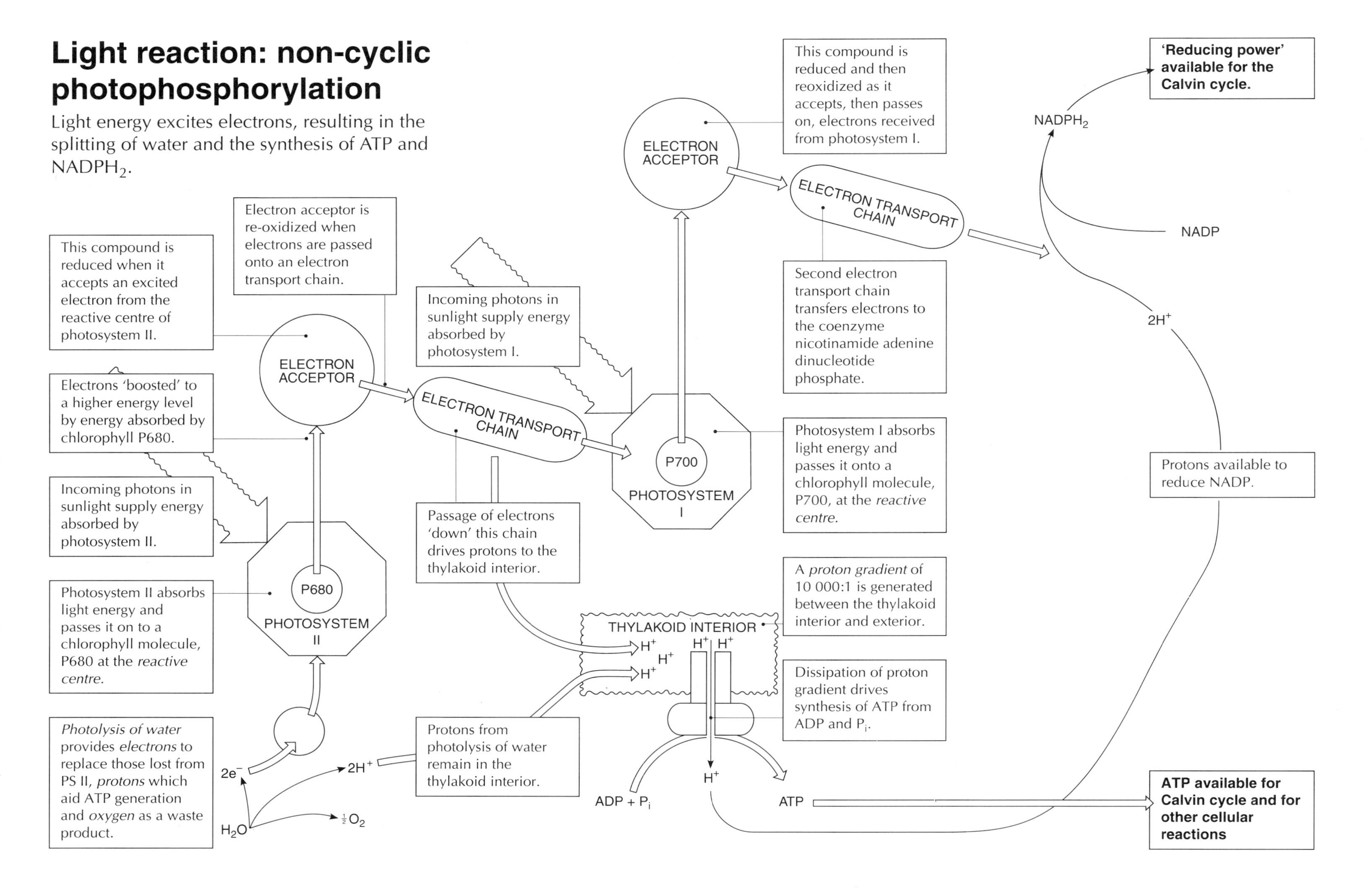

Dark reaction: the Calvin cycle

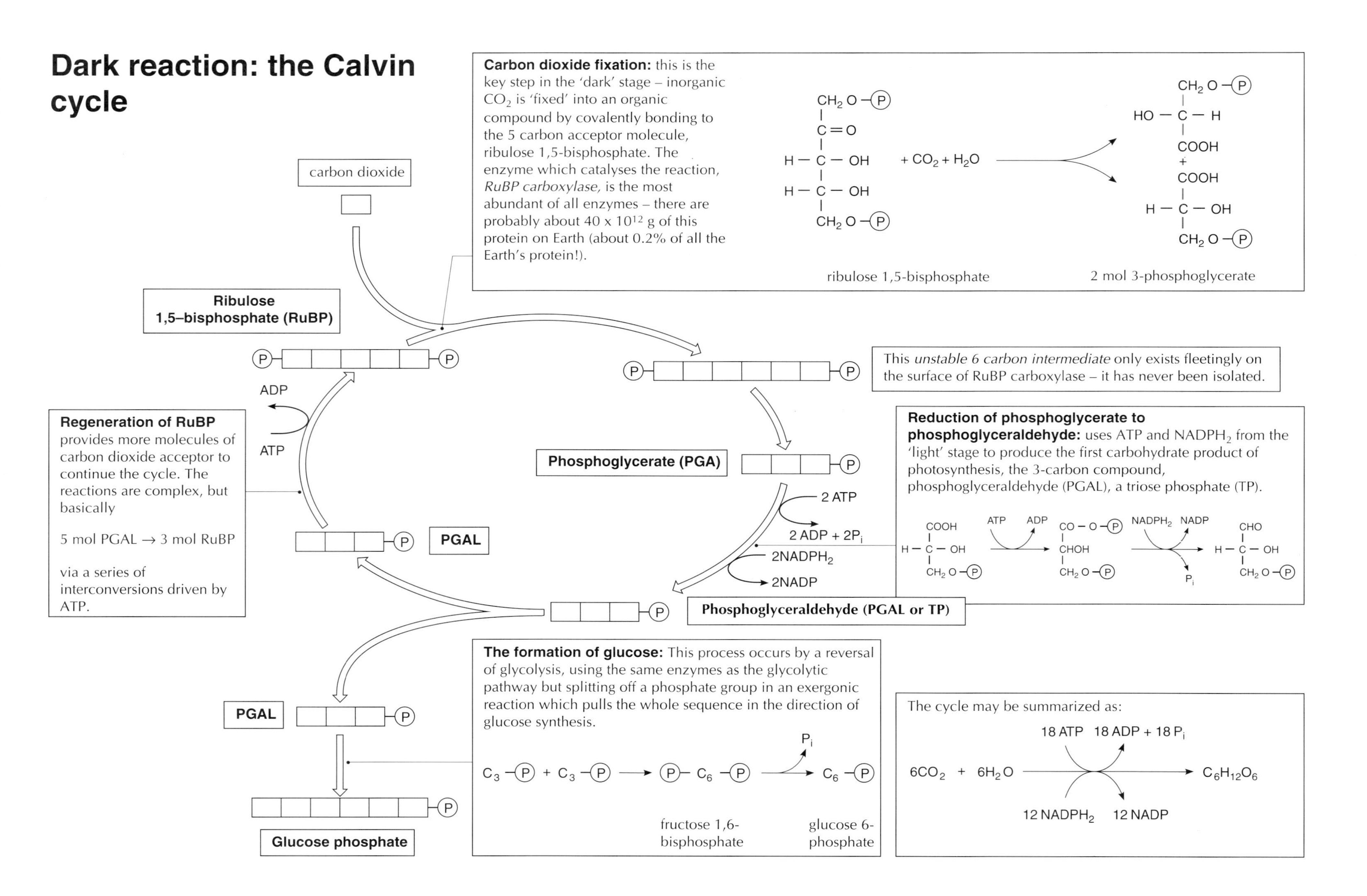

Chloroplasts: absorption and action spectra of chlorophyll

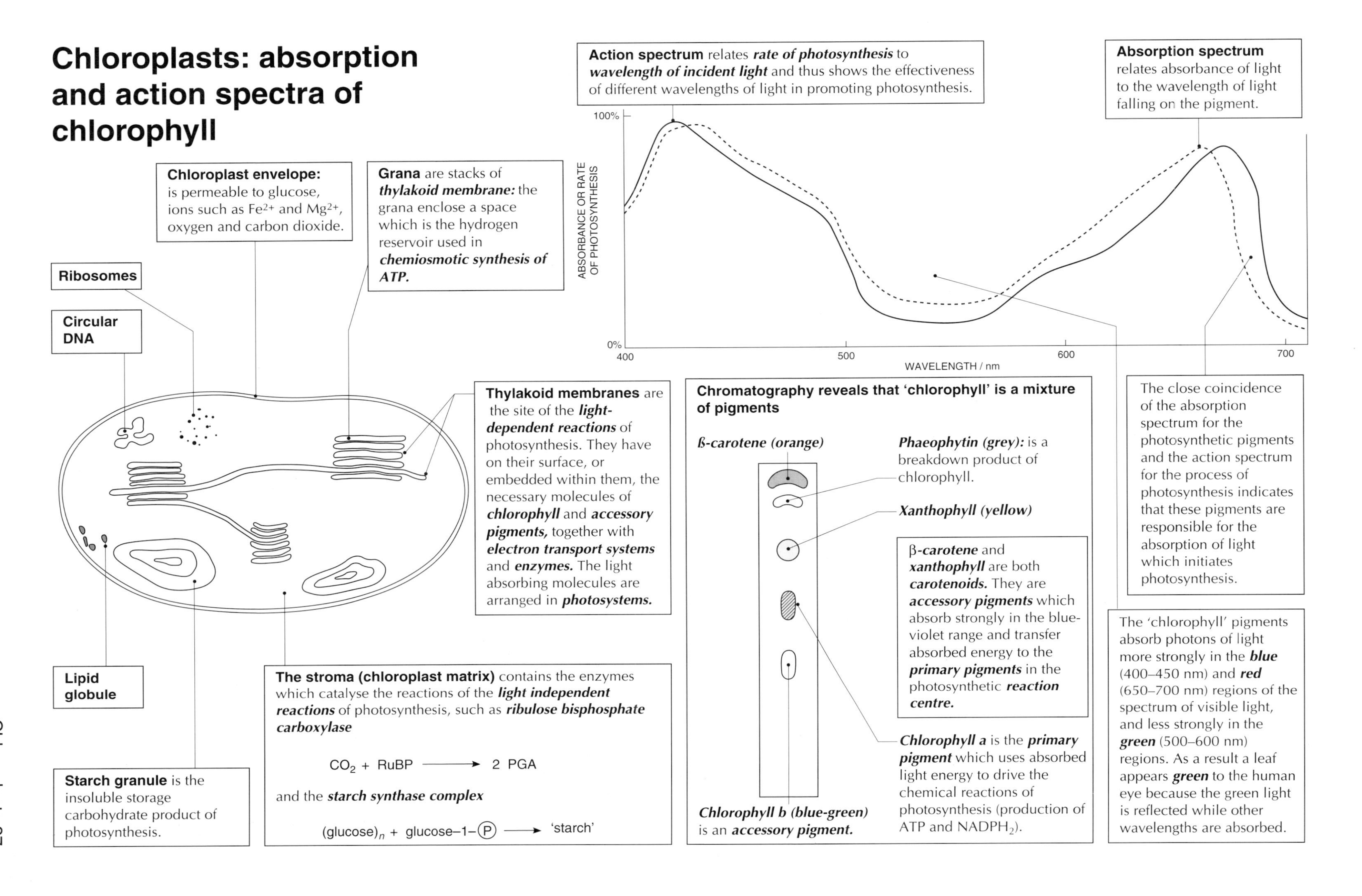

Mineral requirements of plants

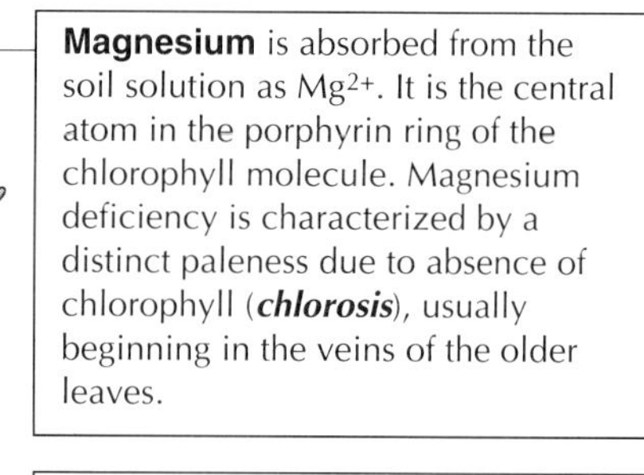

Calcium is absorbed as Ca^{2+}. It forms junctions between the molecules of pectate in the middle lamella, strengthening the binding of adjacent cells to one another. Deficiency leads to die-back of shoots due to death of apical buds.

Phosphorus is absorbed from the soil solution as $H_2PO_4^-$ (a type of phosphate). Its availability is probably ***the major limiting factor*** in plant growth in ***uncultivated soils.*** It is a component of nucleic acids, phospholipids and ATP. Lack of phosphorus usually affects processes which consume ATP, particularly the active uptake of minerals by roots.

Magnesium is absorbed from the soil solution as Mg^{2+}. It is the central atom in the porphyrin ring of the chlorophyll molecule. Magnesium deficiency is characterized by a distinct paleness due to absence of chlorophyll (***chlorosis***), usually beginning in the veins of the older leaves.

Nitrogen is absorbed as NO_3^- or as NH_4^+, and usually converted to amino acids or amides for transport through the plant. Nitrogen is essential for the synthesis of amino acids, proteins, plant hormones, nucleic acids, nucleotides and chlorophyll. Deficiency causes reduced growth of all organs, particularly the leaves, and a marked chlorosis. Nitrogen availability is probably the ***major limiting factor*** in plant growth in ***cultivated soils.***

HYDROPONICS ('growing in water') permits study of plant deficiency symptoms in controlled conditions which eliminate the many variables associated with soil as a growth medium.

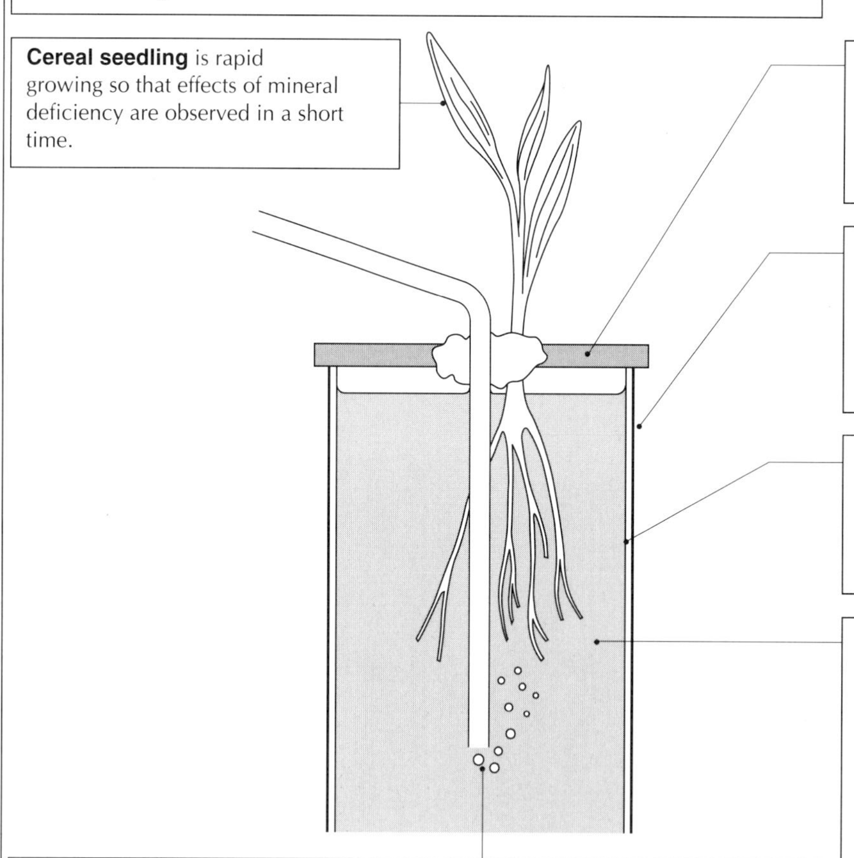

Cereal seedling is rapid growing so that effects of mineral deficiency are observed in a short time.

Lightproof cover
1. prevents entry of airborne contaminants and
2. supports seedling in growing position.

Black card or ***foil cover*** prevents entry of light so that no aquatic photoautotrophs can compete with seedling roots for mineral ions. Cover can be easily removed to examine the growth of the seedling roots.

Glass container can be thoroughly cleaned (using acid) so that no minute traces of mineral remain where they might lead to erroneous results.

Growth/nutrient solution
contains mineral ions in previously determined optimum concentrations. Complete (control) solution contains ***all*** ions, test solutions have a single ion omitted. N.B. to eliminate ***one*** ion, a salt is omitted but its 'second' ion is replaced.
e.g. for nitrogen-free solution omit calcium nitrate but increase concentration of another calcium salt.
The complete solution is often called ***Knop's solution.***

Aeration
has two functions:
1. mixing of solution so that no stagnation occurs;
2. oxygenation so that aerobic (root) respiration may provide energy for active uptake of ions from solution by roots.

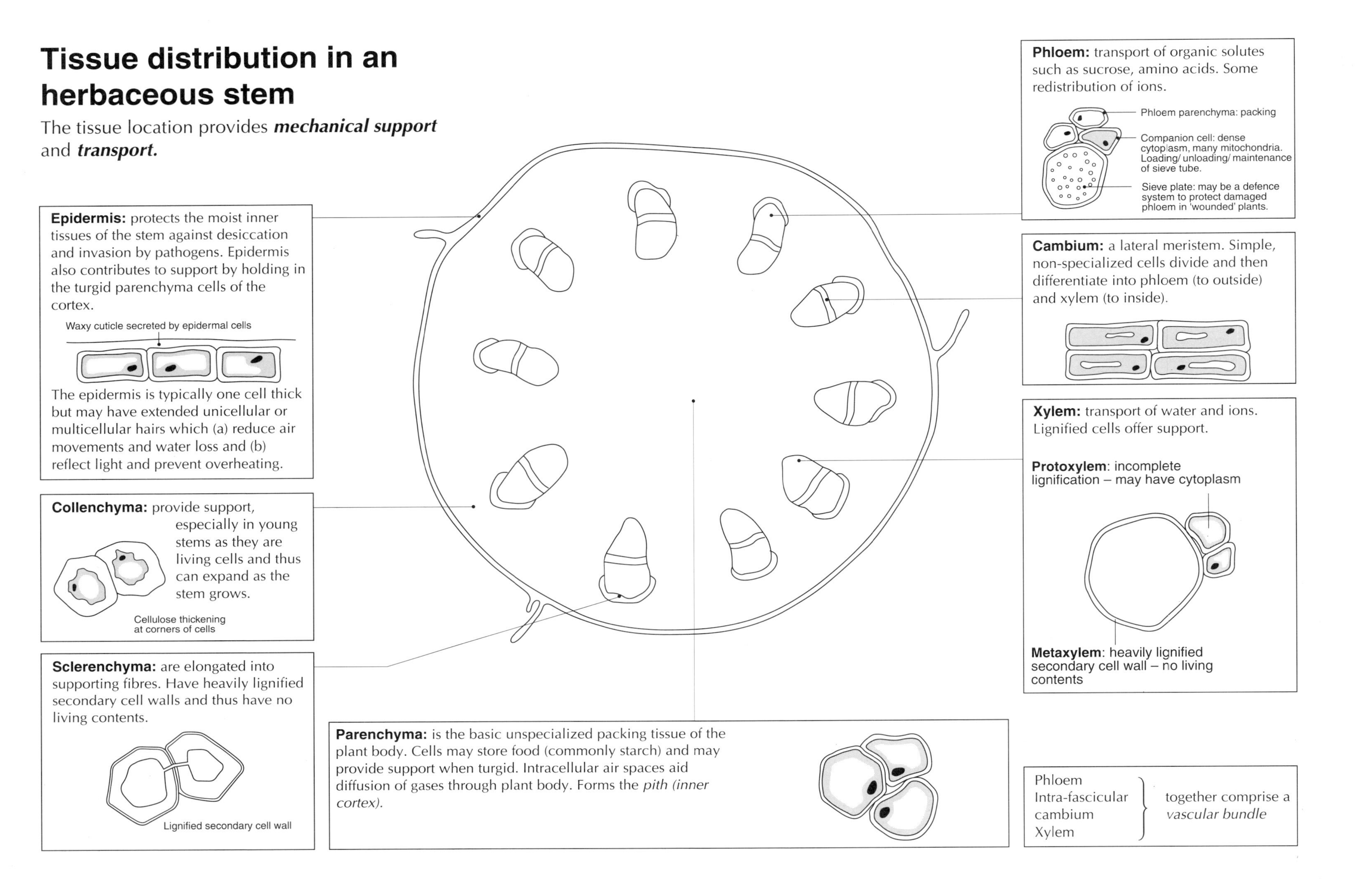
Tissue distribution in an herbaceous stem
The tissue location provides mechanical support and transport.
Epidermis: protects the moist inner tissues of the stem against desiccation and invasion by pathogens. Epidermis also contributes to support by holding in the turgid parenchyma cells of the cortex.
Waxy cuticle secreted by epidermal cells
The epidermis is typically one cell thick but may have extended unicellular or multicellular hairs which (a) reduce air movements and water loss and (b) reflect light and prevent overheating.
Collenchyma: provide support, especially in young stems as they are living cells and thus can expand as the stem grows.
Cellulose thickening at corners of cells
Sclerenchyma: are elongated into supporting fibres. Have heavily lignified secondary cell walls and thus have no living contents.
Lignified secondary cell wall
Parenchyma: is the basic unspecialized packing tissue of the plant body. Cells may store food (commonly starch) and may provide support when turgid. Intracellular air spaces aid diffusion of gases through plant body. Forms the pith (inner cortex).
Phloem: transport of organic solutes such as sucrose, amino acids. Some redistribution of ions.
Phloem parenchyma: packing
Companion cell: dense cytoplasm, many mitochondria. Loading/ unloading/ maintenance of sieve tube.
Sieve plate: may be a defence system to protect damaged phloem in 'wounded' plants.
Cambium: a lateral meristem. Simple, non-specialized cells divide and then differentiate into phloem (to outside) and xylem (to inside).
Xylem: transport of water and ions. Lignified cells offer support.
Protoxylem: incomplete lignification – may have cytoplasm
Metaxylem: heavily lignified secondary cell wall – no living contents
Phloem
Intra-fascicular cambium
Xylem
together comprise a vascular bundle

Tissue distribution in a dicotyledonous root

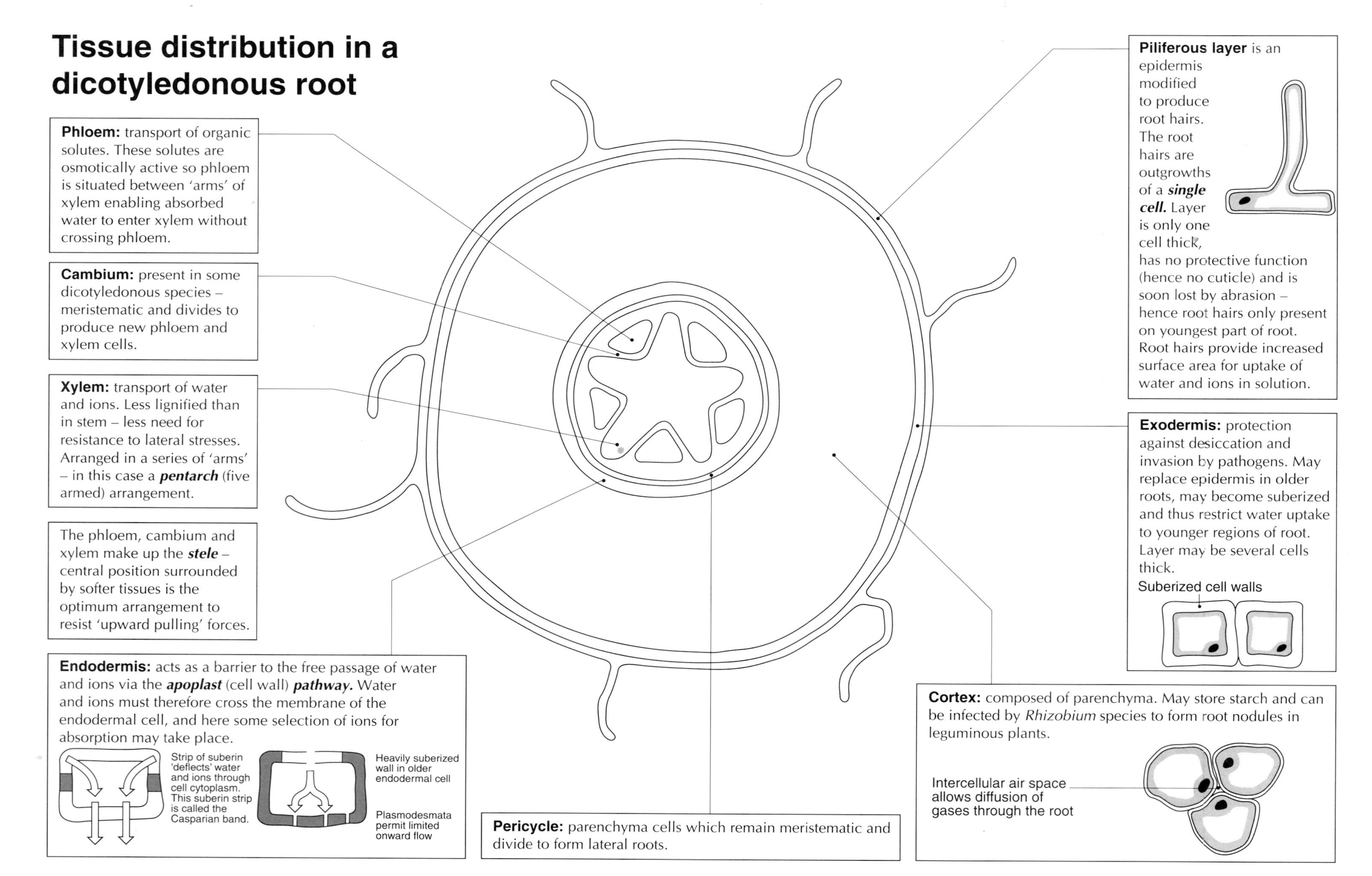

Evidence for phloem as the tissue for translocation comes from the use of radioactive tracers, aphids and metabolic poisons.

The pattern of movement of these solutes within the plant body has also been investigated using radioisotopes, and it has been shown that the pattern of movement may be modified as the plant ages. Up to maturity the lower leaves of an actively photosynthesizing plant may pass their products to the roots for consumption and storage, but once fruit formation begins, ever-increasing numbers of leaves pass their products up to the fruits and eventually even the lower leaves are doing so. Minerals are often remobilized – having been delivered to the photosynthetic leaves via the xylem they may be re-exported through the phloem as the leaves age prior to abscission. The direction of solute movement is under the control of plant growth substances, particularly IAA and the cytokinins.

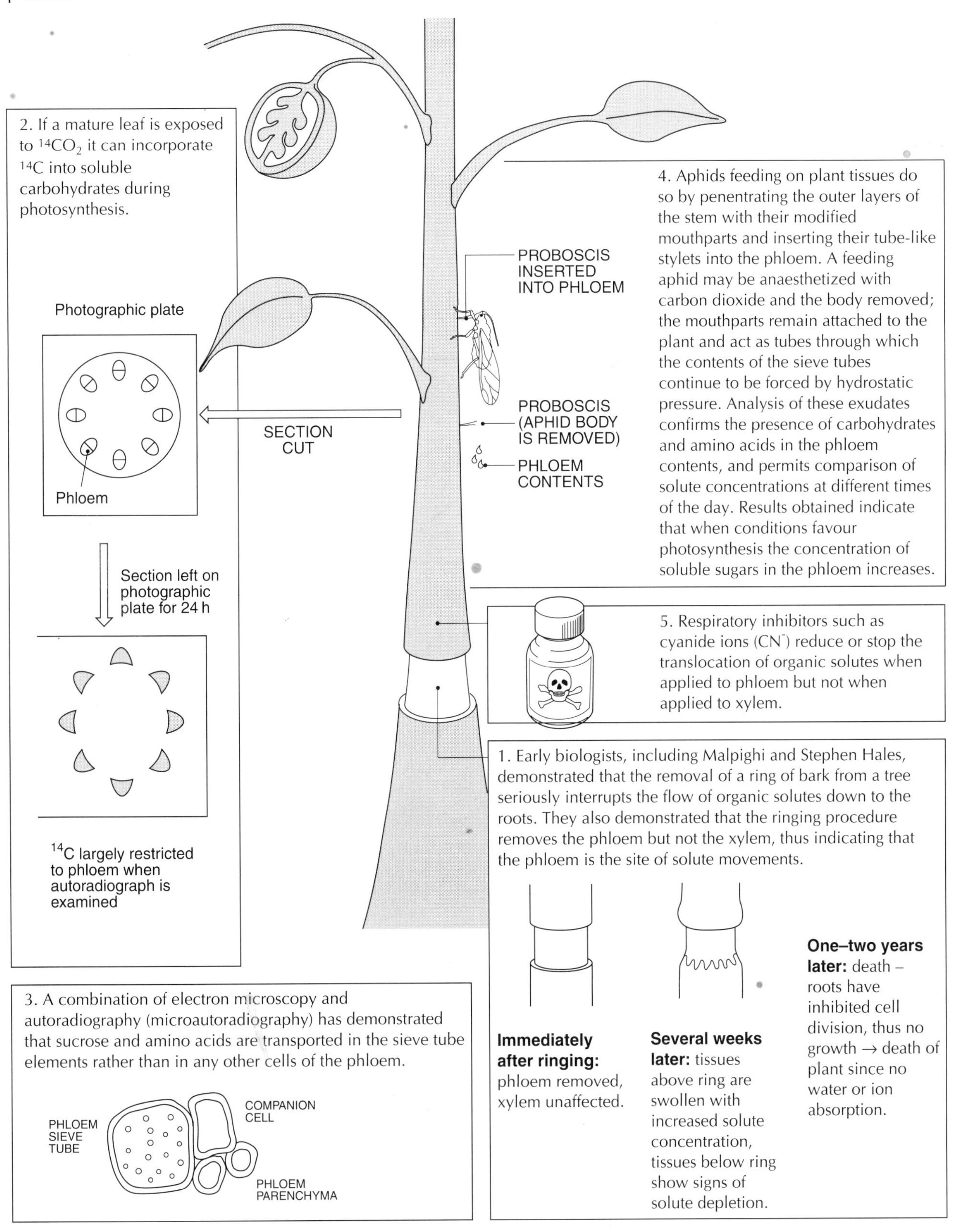

Water potential

Water potential is a measure of the free kinetic energy of water in a system, or the tendency of water to leave a system. It is measured in units of pressure (kPa) and is given the symbol ψ ('psi').

For pure water the water potential is arbitrarily given the value 0: this is a reference point, rather like the redox potential system used in chemistry.

i.e. for pure water $\psi = 0$

In a solution the presence of molecules of solute prevents water molecules leaving. Thus

$$\psi_{SOLUTION} < 0$$

(In the solution the solute molecules 'hinder' the movement of the water molecules, thus the kinetic energy of the water molecule is reduced and ψ becomes negative.)

Water moves down a gradient of water potential, i.e. from a less negative (e.g. -500 kPa) to a more negative (e.g. -1000 kPa) water potential.

WATER POTENTIAL OF SYSTEM kPa

0

–500

–1000

The advantages of the water potential nomenclature

1. the movement of water is considered from the 'system's' point of view, rather than from that of the environment;
2. comparison between different systems can be made, e.g. between the atmosphere, the air in the spaces of a leaf and the leaf mesophyll cells.

We should remember that:

Osmosis is the movement of water, through a partially permeable membrane, along a water potential gradient.

The presence of solutes in the cell sap makes ψ_s lower (more negative). The ψ for pure water = 0, so a plant cell placed in pure water will experience an inward movement of water, by osmosis, along a water potential gradient.

The cellulose cell wall is freely permeable to water and to solutes.

The plasma membrane is freely permeable to water but of limited permeability to solutes.

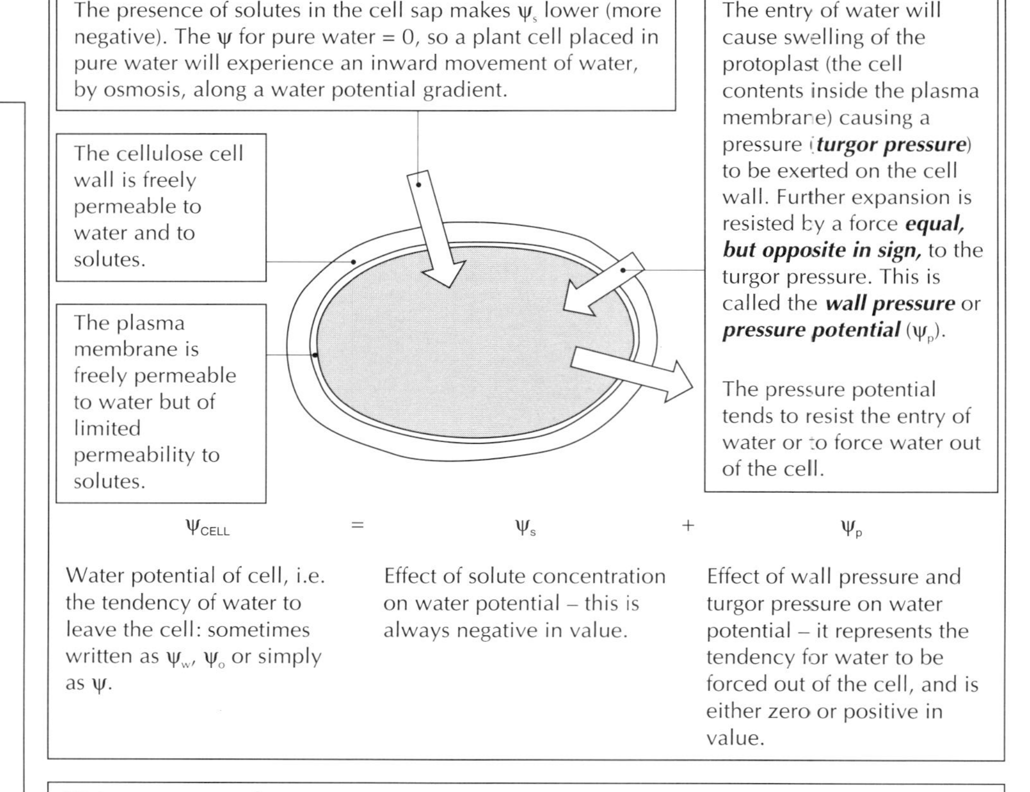

The entry of water will cause swelling of the protoplast (the cell contents inside the plasma membrane) causing a pressure (***turgor pressure***) to be exerted on the cell wall. Further expansion is resisted by a force ***equal, but opposite in sign,*** to the turgor pressure. This is called the ***wall pressure*** or ***pressure potential*** (ψ_p).

The pressure potential tends to resist the entry of water or to force water out of the cell.

$$\psi_{CELL} = \psi_s + \psi_p$$

ψ_{CELL}: Water potential of cell, i.e. the tendency of water to leave the cell: sometimes written as ψ_w, ψ_o or simply as ψ.

ψ_s: Effect of solute concentration on water potential – this is always negative in value.

ψ_p: Effect of wall pressure and turgor pressure on water potential – it represents the tendency for water to be forced out of the cell, and is either zero or positive in value.

Water movement between cells:

1. Calculate ψ for each cell from ψ_s and ψ_p.
2. Predict direction of water movement since water moves ***down the gradient of water potential.***

Cell A Cell B

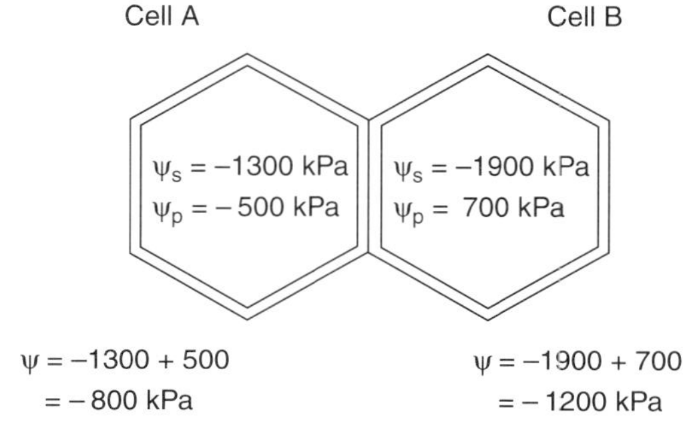

Cell A: $\psi = -1300 + 500 = -800$ kPa

Cell B: $\psi = -1900 + 700 = -1200$ kPa

Since -800 is a higher number than -1200 ***water will move by osmosis down a water potential gradient from A → B.***

Water relationships of plant cells

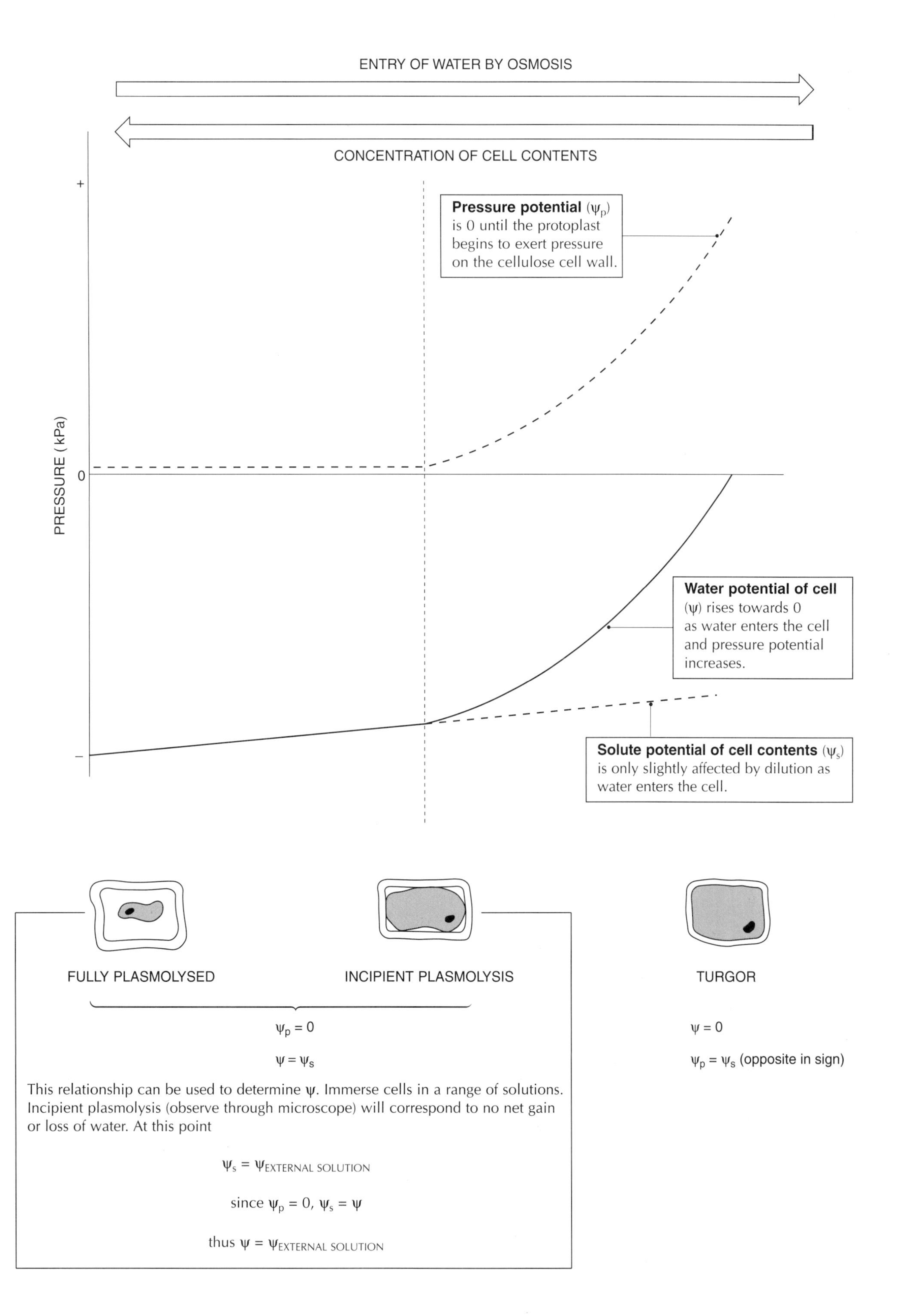

This relationship can be used to determine ψ. Immerse cells in a range of solutions. Incipient plasmolysis (observe through microscope) will correspond to no net gain or loss of water. At this point

$$\psi_s = \psi_{\text{EXTERNAL SOLUTION}}$$

$$\text{since } \psi_p = 0,\ \psi_s = \psi$$

$$\text{thus } \psi = \psi_{\text{EXTERNAL SOLUTION}}$$

Stomata represent an important adaptation to life in a terrestrial environment.

Guard cells

1. Change *shape* as their *degree of turgor* is altered. The reason for this is twofold:
   a. the inner guard cell wall has microfibrils orientated so that longitudinal expansion is not easy;
   b. the outer wall is less thickened with cellulose so that it elongates much more readily than the inner wall.

2. May alter their *solute potential,* and thus their *water potential,* by the movement of ions, principally potassium ions, K^+.

An ATP-dependent proton pump moves H^+ *out of* the guard cells, creating an electrochemical gradient (inside *negative* with respect to *outside*) so that K^+ can flow passively through K^+ channels down this electrochemical gradient.

K^+ movement *in* → reduced water potential → water movement *in.*

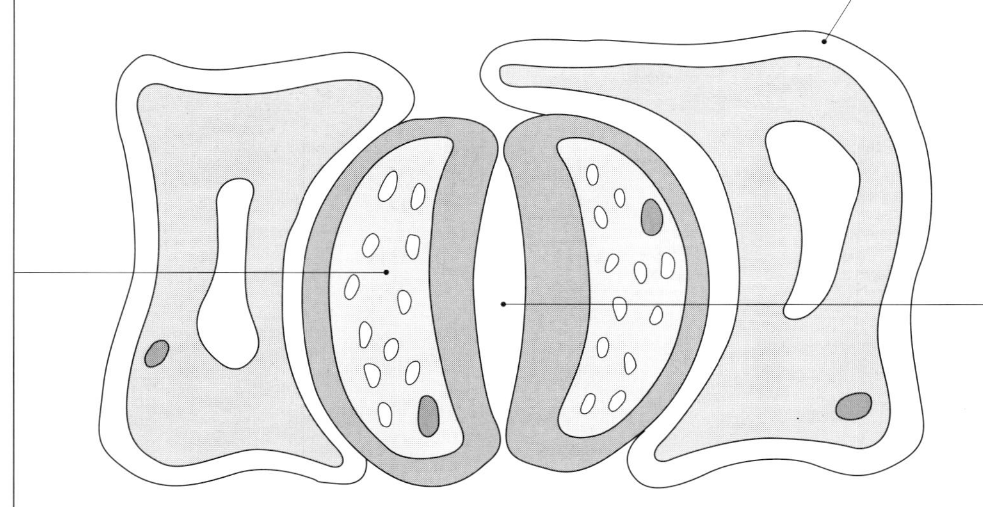

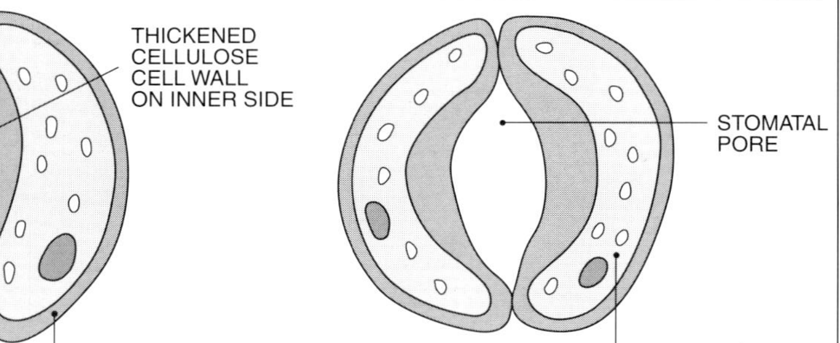

Closed: guard cells are *flaccid* because water has been lost to the *apoplast system* of the epidermis.

Reason: $[K^+]$ in guard cells has fallen – high water potential thus allows water to leave the guard cells.

Open: Guard cells are *turgid* because water has been gained from the *apoplast system* of the epidermis.

Reason: $[K^+]$ in guard cells has risen – low water potential thus allows water to enter the guard cells.

Adjacent epidermal cells: note that there is no *symplastic* connection (i.e. no *plasmodesmata*) between these cells and the guard cells. The adjacent epidermal cells do not have the chloroplasts or the dense cytoplasm typical of guard cells.

Control of stomatal aperture: must permit
a. entry of enough CO_2 to permit photosynthesis;
b. control water loss to prevent desiccation of plant tissues

and has both *internal* (plant growth regulators) and *environmental* (e.g. air humidity) *signals.*

Stomatal opening is promoted by:

1. *low intercellular* ***[CO₂]****:* sensed by guard cells and corresponds to the need for more CO_2 to maintain rates of photosynthesis.
2. *high light intensity:* light absorbed by chlorophyll (the PAR photosystem) provides ATP by photophosphorylation – this increases proton pumping and 'opens' stomata when light is available – 'anticipating' the need for more CO_2 for photosynthesis. This is a *feedforward response.* There is a second (blue-light photosystem) response for opening stomata in shady conditions, or at dawn.

Stomatal closure is triggered by:

1. *low environmental humidity;*
2. *increasing leaf temperature;*

   both of which are signals that the need to conserve water must override the need to allow CO_2 uptake. This response is triggered by the water content of leaf epidermal and mesophyll cells and it is a *feedback response.* (N.B. at very high temperatures stomata may open very wide to allow maximum leaf cooling by evapotranspiration from mesophyll cells.)
3. *Abscisic acid secretion:* severe drought stress is detected in epidermal cells which secrete ABA into the apoplast causing rapid and immediate stomatal closure (possibly by inhibiting the proton pump).

Cohesion–tension theory of transpiration

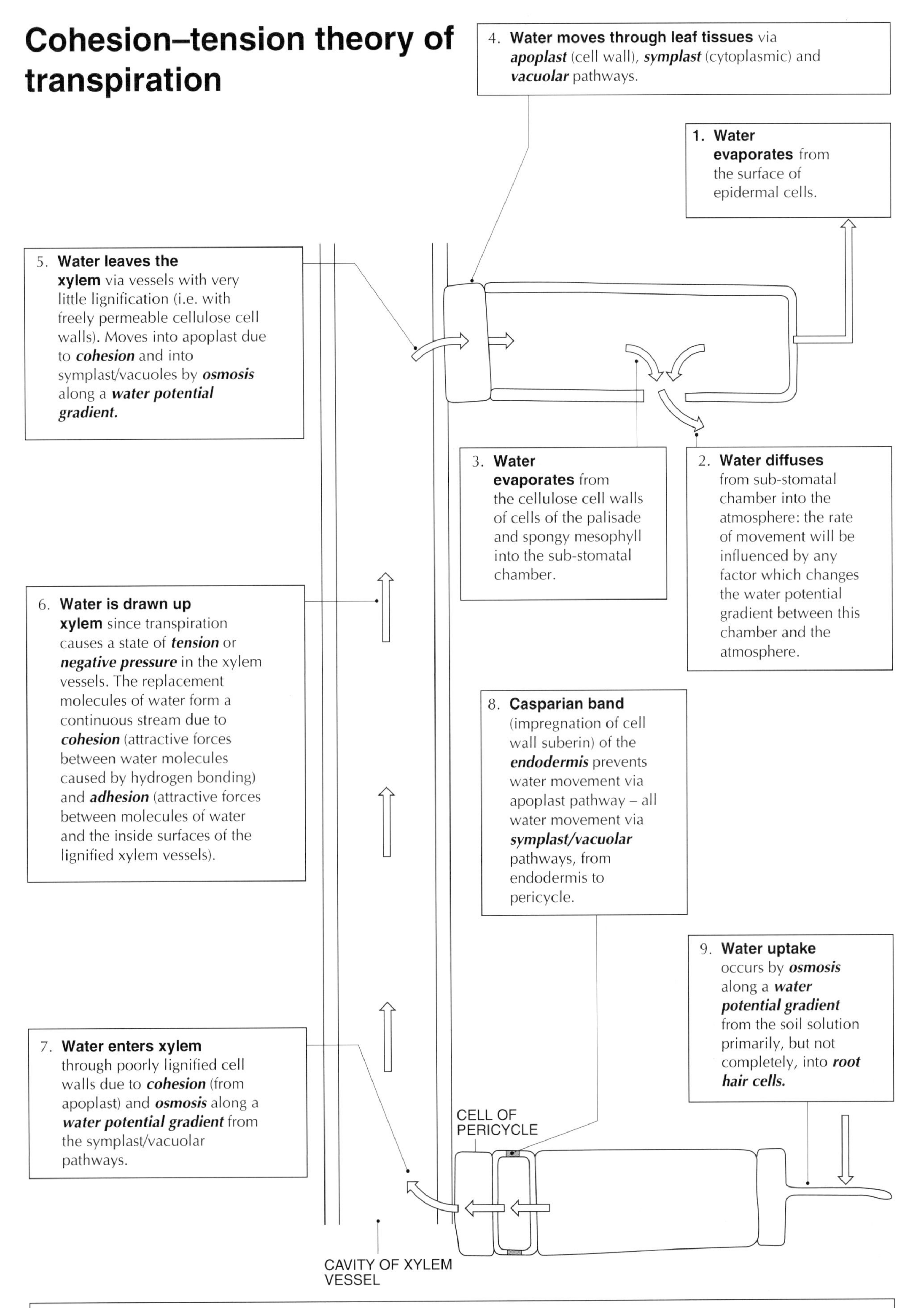

Water movement through the plant occurs as a result of very large differences in water potential between the atmosphere and the soil solution. The process begins with ***evaporation*** from the leaf surfaces, is continued due to ***cohesion*** between water molecules and ***tension*** in the xylem vessels and is completed by ***osmosis*** from the soil solution.

The bubble potometer

measures ***water uptake*** (= water loss by transpiration + water consumption for cell expansion and photosynthesis).

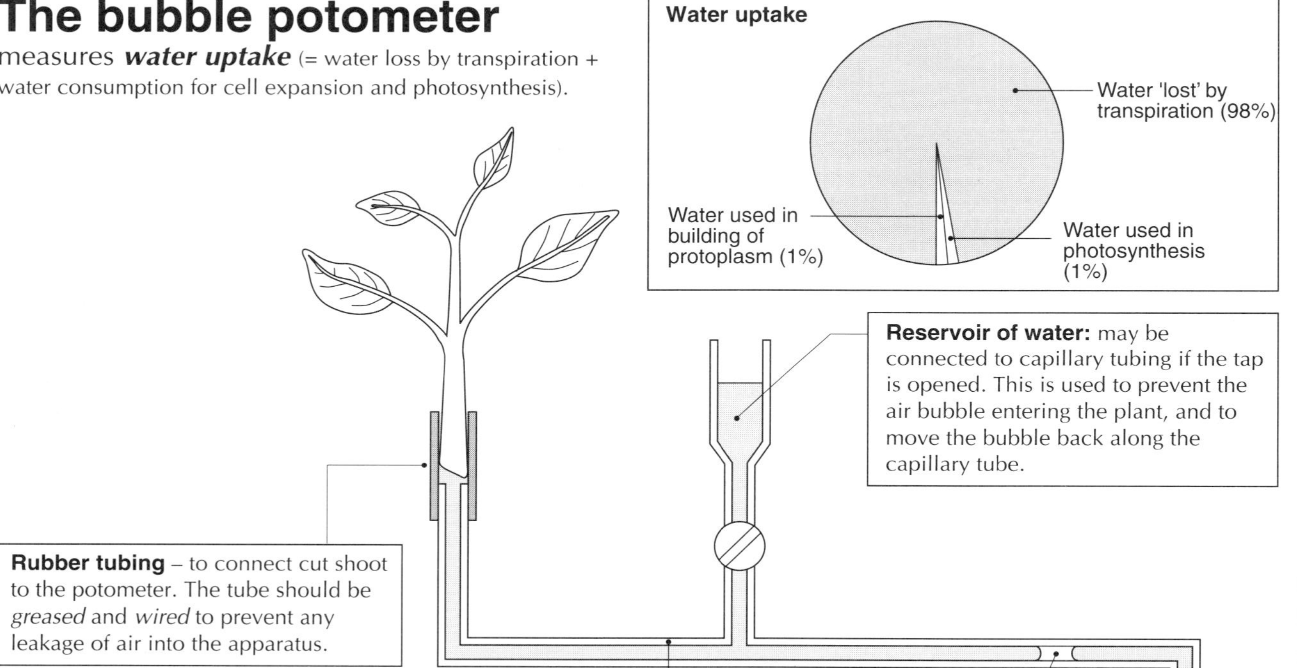

Reservoir of water: may be connected to capillary tubing if the tap is opened. This is used to prevent the air bubble entering the plant, and to move the bubble back along the capillary tube.

Rubber tubing – to connect cut shoot to the potometer. The tube should be *greased* and *wired* to prevent any leakage of air into the apparatus.

Capillary tube: must be kept horizontal to prevent the bubble moving due to its density compared with water.

Air bubble: inserted by removal of tube end from beaker of water. Movement corresponds to water uptake by the cut shoot.

Graduated scale: permits direct reading of bubble movement/water uptake.

Atmometer control: The atmometer is an instrument which can measure evaporation from a non-living surface. When subjected to the same conditions as a potometer the changes in the rate of evaporation from a plant and from a purely physical system can be compared – for example, a reduction in light intensity will show a decrease in water loss *only from a potometer* (due to stomatal closure). The atmometer control indicates when the potometer is acting as a free evaporator and when it is affected by physiological factors such as photosynthesis and stomatal closure.

PROCEDURE

1. The leafy shoot must be cut *under water*, the apparatus must be filled *under water* and the shoot fixed to the potometer *under water* to prevent air locks in the system.
2. Allow plant to equilibrate (5 min) before introduction of air bubble. Take at least three readings of rate of bubble movement, and use reservoir to return bubble to zero on each occasion. Calculate mean of readings. Record air temperature.
3. Scale can be calibrated by introducing a known mass of mercury into the capillary tubing and using $\rho = m/v$ (ρ for mercury is known, m can be measured, thus v corresponding to a measured distance of bubble movement can be determined).
4. Rate of water uptake per unit area of leaves can be calculated by measurement of leaf area.

EXTERNAL FACTORS AFFECTING TRANSPIRATION

Light intensity: use bench lamp (with water bath to act as heat filter) to increase light intensity. To simulate 'darkness' enclose shoot in black polythene bag.

Wind: use small electric fan with 'cool' control to mimic air movements whilst avoiding effects of temperature changes.

Humidity: enclose shoot in clear plastic bag to *increase* relative humidity of atmosphere – include water absorbant such as calcium chloride to *decrease* relative humidity.

May also determine relative importance of upper surface/lower surface/stem/petiole in water loss by smearing with vaseline (acts like a waxy cuticle) as appropriate.

N.B. It is sometimes difficult to change only one condition at a time, e.g. enclosure in a black bag to eliminate light will also increase the relative humidity of the atmosphere.

Plant growth substances

Auxins are used as defoliants, e.g. during the Vietnam War to clear areas of vegetation and make bombing of bridges, roads, and troops easier. Also used to remove vegetation from overhead power lines – manual removal would be costly and dangerous.

A mixture of ***auxin, cytokinin*** and ***gibberellin*** will inhibit apical growth and allow limited development of lateral buds. This mixture applied to hedges promotes dense, bushy growth and limits the need for mechanical trimming to one or two occasions per year.

Gibberellic acid may mimic red light: control of flowering time (promote long-day species, inhibit short-day species) means flowers can be available 'out of season'.

Ethene sprayed onto day-neutral species such as pineapple can synchronize flowering/fruiting so that crop picking can be more efficient.

Auxins used as pre-emergent herbicides to prevent germination of weed species and as post-emergent herbicides to remove scrub or competitors for crops.

Cytokinins delay leaf senescence and are used to maintain the life of fresh, leafy crops such as lettuce.

Auxin can inhibit 'sprouting' (lateral bud development) in stored potatoes.

NATURAL PLANT GROWTH HORMONES

Growth of stem: cell enlargement is promoted by ***auxin*** and ***gibberellin.*** Redistribution of ***auxin*** causes phototropism.

Seed dormancy is maintained by ***abscisic acid*** but is broken by ***gibberellic acid.***

Leaf fall is promoted by ***abscisic acid.***

Root growth is ***promoted*** by ***auxin*** at ***low*** concentration but ***inhibited*** at ***high auxin*** concentration.

Lateral bud development is inhibited by ***auxin*** but promoted by ***cytokinin (antagonism).***

Stomatal closure under stress may be promoted by ***abscisic acid.***

Flowering may be triggered by ***florigen.***

Root growth of adventitious roots is promoted by ***auxin.***

N.B. Many commercial applications of these growth phenomena rely on ***plant growth regulators,*** which are synthetic derivatives of the natural compounds, but are usually more effective in lower concentrations because they are degraded less rapidly by the plant.

Cytokinin can promote fruit growth, and synergistically with ***auxin*** and ***gibberellin*** can promote parthenocarpy. This is useful if seed fails to 'set' due to poor pollinating conditions, and 'seedless' (parthenocarpic) fruits are popular with consumers.

Gibberellins increase fruit size in grapes if applied just after flowers open since some ovules abort – 'crowding' is reduced allowing more nutrients to reach remaining fruits and limiting fungal infections.

Auxin can prevent premature fruit drop (windfall losses) since it is antagonistic to ***abscisic acid.***

Auxin can act as a selective lawn weed killer since broad leaved 'weed' species are killed by auxin concentrations which do not affect monocotyledons.

Ethene is used to accelerate ripening – ideal for grapes which can be picked earlier and thus have a longer drying period for forming raisins. Ripening can be delayed by keeping fruits in an oxygen-free atmosphere: ethene can then induce ripening as required.

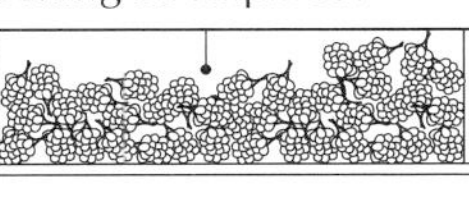

Structure of a typical flower

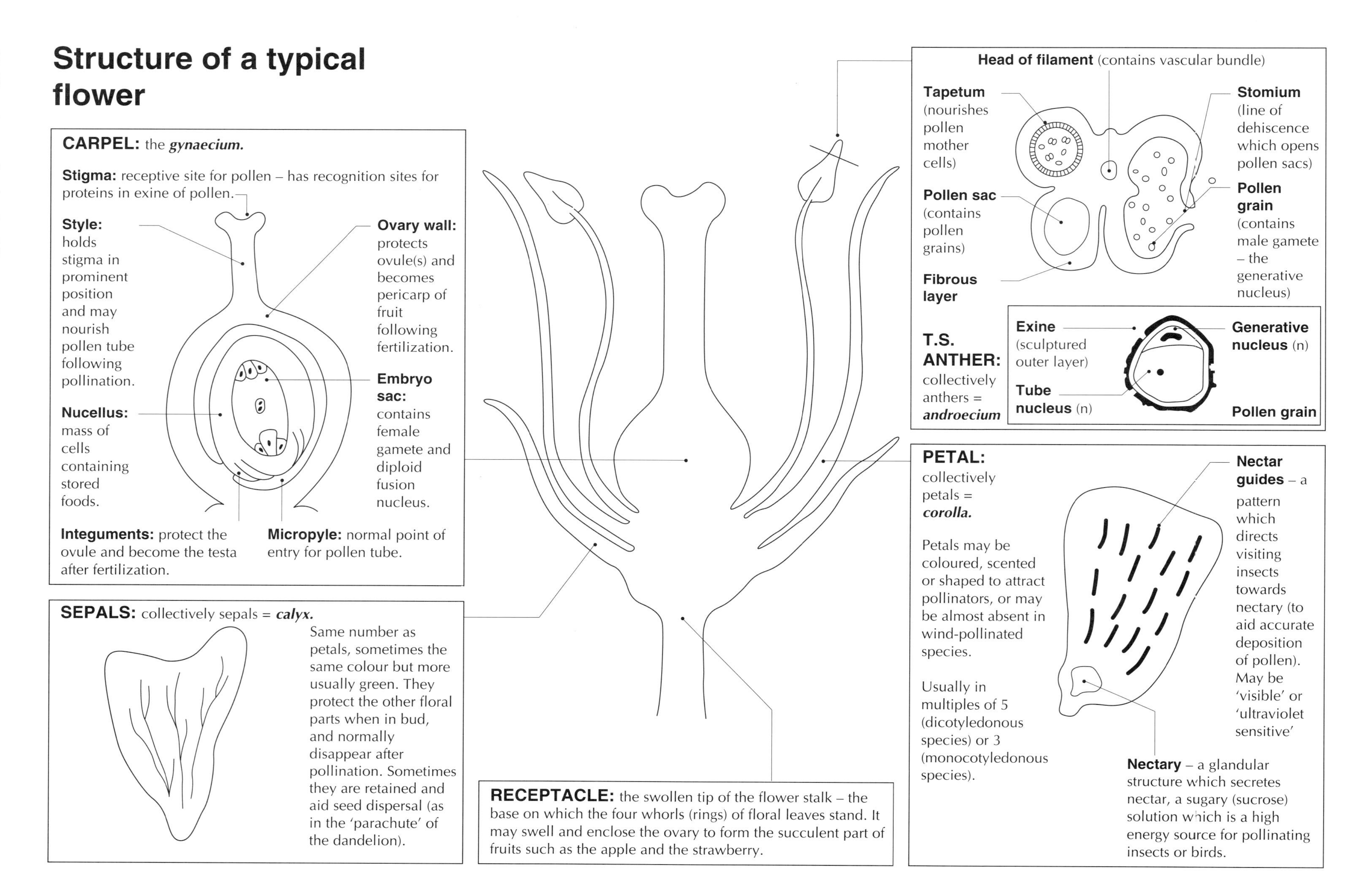

Wind-pollinated (anemophilous) flowers

are typically grasses or forest tree species which occur in dense groups covering very large areas.

Bract: leaf-like structure which encloses and protects floral structures.

Filaments are long and flexible so that anthers may be held out in an exposed position.

Anthers are versatile – hinged at the mid-point – so that pollen is readily shaken out of flower by the wind.

Pollen is produced in enormous quantities since transfer by wind is very wasteful. The pollen is smooth, with wing-like extensions, light and small to promote transfer by wind.

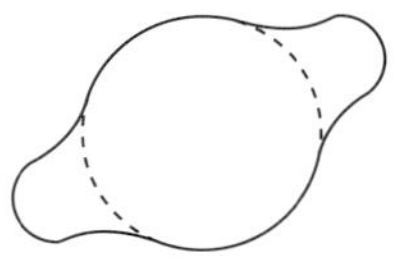

Stigmas are long and feathery giving a large surface area to receive pollen, and often protrude outside the flower into the pollen-bearing atmosphere.

Petals are dull in colour and much reduced in size, since they need not attract insects and must not obstruct pollen access to stigma. Together with the very small *sepals* the petals form the *lodicule* which swells to open the flower by forcing apart two *bracts*.

Neither nectar nor scent is produced: the production of these compounds would be biochemically 'expensive' and would offer no advantage since there is no need to attract pollinating insects.

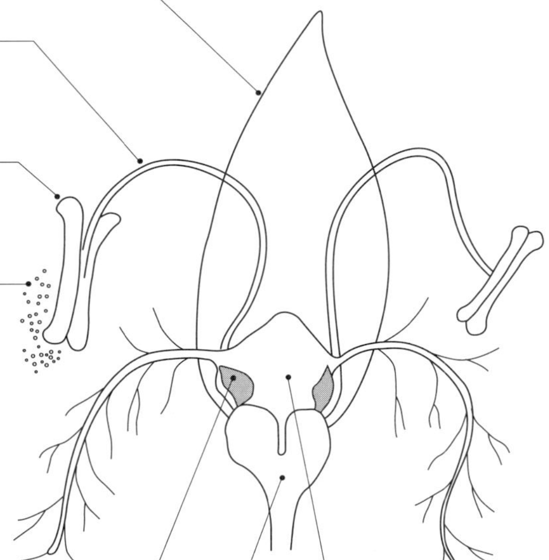

Receptacle

Ovary contains a single ovule.

Insect-pollinated (entomophilous) flowers

typically belong to species which are solitary or exist in small groups.

Filaments are stiff to resist buffeting by pollinating insects, short enough to keep anther within the corolla and may be hinged to aid deposition of pollen on body of visiting insect.

Anthers are held within corolla and may have a sticky surface to hold pollen ready for visiting insect.

Pollen is large with sticky projections to adhere to body of insect. Less pollen is produced as transfer by insect is very efficient.

Stigma is held within the corolla in a position which ensures contact with body of visiting insect. Stigma surface is relatively small, since insect transfer is accurate, and often the style is lignified/stiffened to avoid damage by pollinating insect.

Petals are brightly coloured and large. They must attract pollinating insects and often have markings or hairs to act as *nectar guides* (these may be visible in the ultraviolet light to which insects are sensitive). Petals may be shaped to attract pollinators (by mimicry of female insects in some orchids) or may be reinforced to act as *landing platforms* for pollinators or as *tunnels* to direct insects towards the reproductive parts.

Scent and/or nectar may be produced to attract or reward visiting insects. Scents are often derivatives of fatty acids and nectar is a dilute sugar (sucrose/fructose) solution with a high energy value.

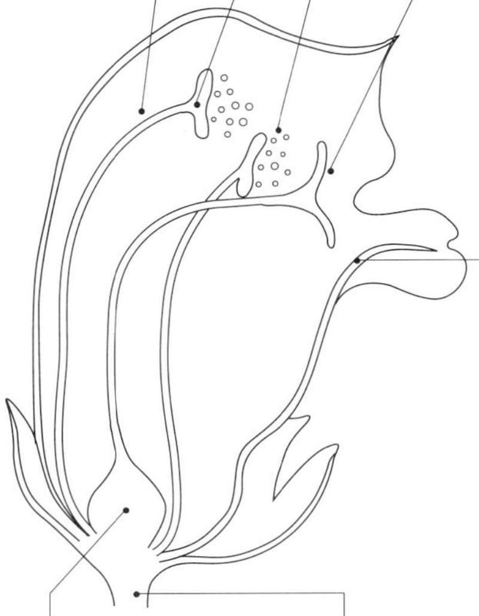

Ovary contains ovules.

Receptacle

The seed is a fertilized ovule

1. It has the advantages of genetic variation.
2. It can remain dormant and survive adverse conditions.
3. It contains a food store for the developing embryo.

LIFE CYCLE OF A FLOWERING PLANT

involves ***meiosis*** to produce ***spores*** and a unique ***double fertilization.***

Embryo (2n) contained within ***testa*** is the ***seed.***

GERMINATION, GROWTH AND DEVELOPMENT

Mature sporophyte: typical flowering plant. All cells are ***diploid*** (2n).

Flower: the most obvious of all the adaptations of the flowering plants. Four whorls (rings) of leaves adapted to produce spores and allow fusion of their products. Cells are ***diploid***

MEIOSIS IN POLLEN SAC OF ANTHER

MEIOSIS IN OVULE OF OVARY

Microspores (unripe pollen grains) which are ***haploid*** (n).

Megaspores (unripe embryo sacs) which are ***haploid*** (n).

Microgametophyte (lit. 'small gamete producing plant') is the ***ripe pollen grain*** which is ***haploid*** (n).

Megagametophyte is the ***ripe embryosac*** which is ***haploid*** (n).

Male gamete is the ***generative nucleus*** (n) within the pollen grain.

Diploid fusion nucleus will become ***endosperm*** following fertilization.

Female gamete is the ***oosphere*** (n) contained within the embryo sac awaiting fertilization.

Unique ***double fertilization*** produces ***diploid zygote*** (2n) and ***triploid endosperm nucleus*** (3n).

Zygote (2n) undergoes ***mitosis*** to produce ***embryo.***

SEED GERMINATION

is regulated by both ***internal*** and ***environmental factors.***

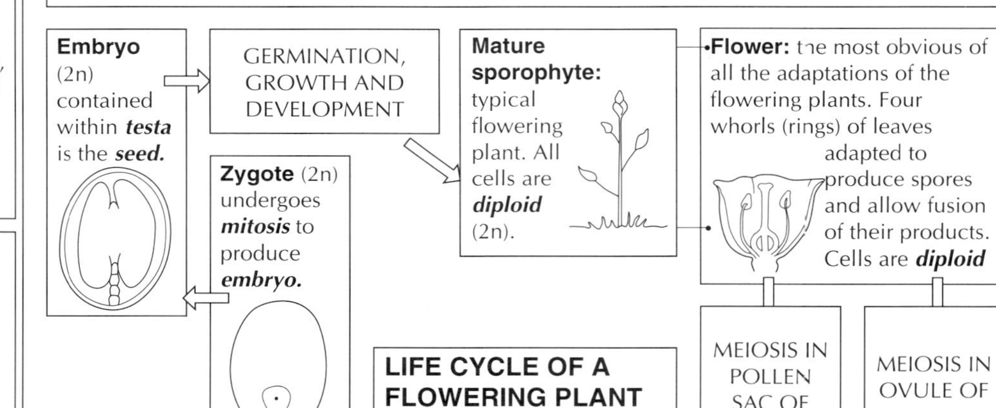

Endosperm contains ***insoluble food stores*** (particularly starch in cereals) which are hydrolysed by enzymes:

STARCH —Amylase→ MALTOSE —Maltase→ GLUCOSE

PROTEINS —Protease→ AMINO ACIDS

LIPIDS —Lipase→ FATTY ACIDS + GLYCEROL

Aleurone layer is three cells thick in cereals: contains protein – may be inactive enzymes or may be source of amino acids for synthesis of enzymes under influence of GA.

Hydrolytic enzymes are released from ***aleurone layer*** (in cereals) or from ***lysosomes*** (in other seeds).

Embryo: manufactures and secretes ***gibberellic acid.*** Other hormones, notably ***cytokinin*** and ***indole acetic acid*** promote cell division at the growing apices of the embryo.

Gibberellic acid is secreted from the embryo to the aleurone layer where it promotes the mobilization of amino acids and the condensation of these amino acids into ***hydrolytic enzymes*** such as ***α-amylase*** and a ***protease.***

Scutellum: modified cotyledon which functions as an ***absorptive organ:*** transfer of the soluble products of hydrolysis of food store in endosperm to developing embryo.

Temperature affects germination because of its influence on the rate of enzyme-catalysed reactions, so that the optimum temperature for germination is commonly in the range 20-40 °C. Many seeds require a period of low temperature - ***stratification*** - before germination will begin. The low temperature may ***increase permeability of the testa*** or may reduce the concentration of ***inhibitory compounds*** such as ***abscisic acid.***

Water is required as a reagent in ***hydrolysis of food stores,*** in the ***mobilization of enzymes*** and in the ***transport of the products of hydrolysis.*** It also causes the ***rupture of the testa*** by the swelling of colloidal substances such as ***proteins*** and ***cell wall materials*** as water is absorbed by imbibition through the micropyle and the testa.

Oxygen is required for ***aerobic respiration.*** The transport of nutrients and processes of cell division are very energy demanding - aerobic respiration represents the most efficient energy release from oxidizable food stores. Measurement of ***respiratory quotient*** (= CO_2 released/O_2 consumed) may indicate which food is being used as a respiratory substrate.

Aerobic carbohydrate respiration RQ = 1.0

Aerobic fat respiration RQ = 0.7

Fat carbohydrate conversion (in lipid-rich seeds) RQ = 0.4

Light may be necessary to ***break seed dormancy.*** Light of the appropriate wavelength is absorbed by ***phytochrome*** and raises the ***gibberellic acid: abscisic acid*** ratio.

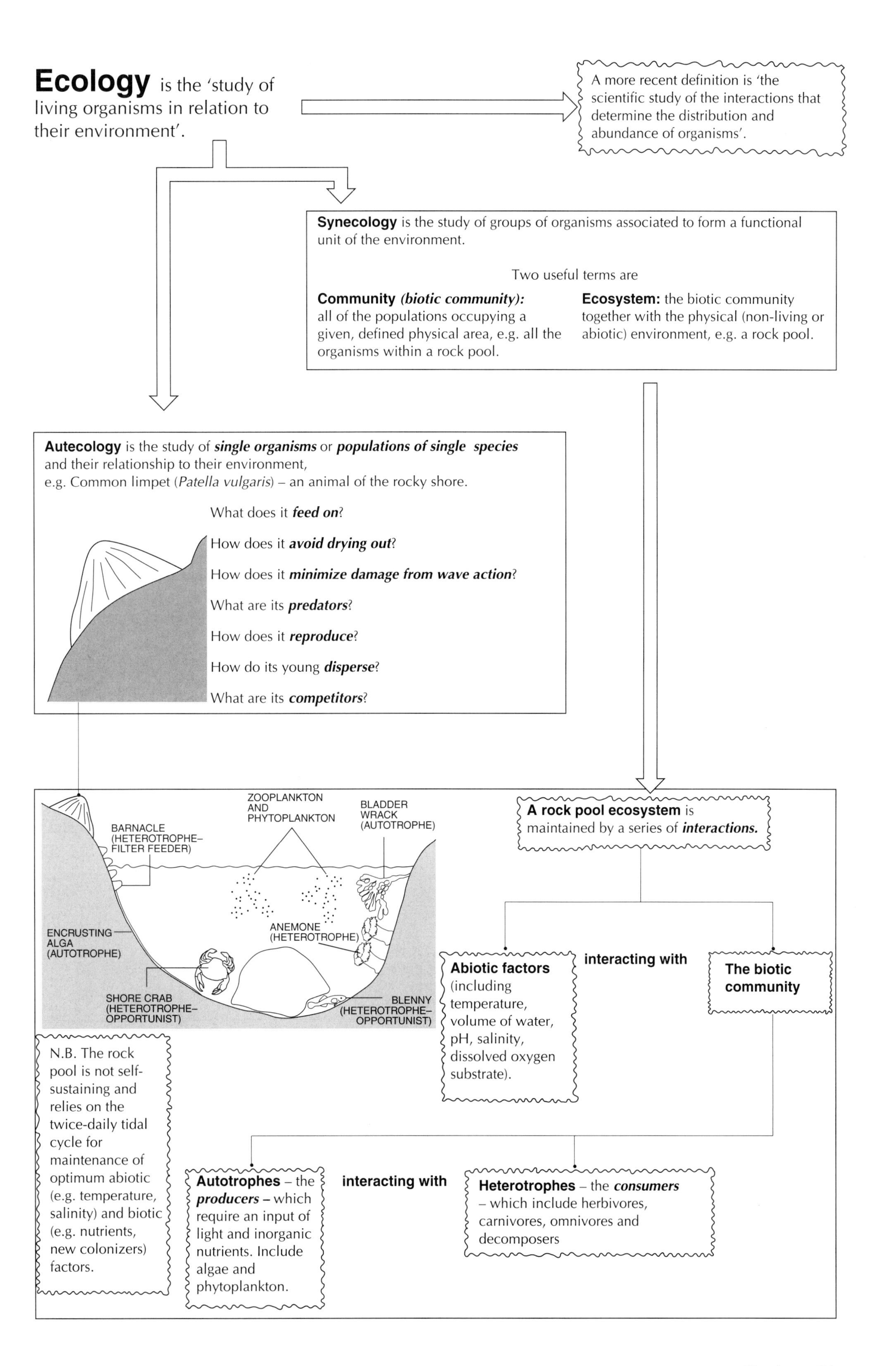
Ecology is the 'study of living organisms in relation to their environment'.
A more recent definition is 'the scientific study of the interactions that determine the distribution and abundance of organisms'.
Synecology is the study of groups of organisms associated to form a functional unit of the environment.
Two useful terms are
Community (biotic community): all of the populations occupying a given, defined physical area, e.g. all the organisms within a rock pool.
Ecosystem: the biotic community together with the physical (non-living or abiotic) environment, e.g. a rock pool.
Autecology is the study of single organisms or populations of single species and their relationship to their environment, e.g. Common limpet (Patella vulgaris) – an animal of the rocky shore.
What does it feed on?
How does it avoid drying out?
How does it minimize damage from wave action?
What are its predators?
How does it reproduce?
How do its young disperse?
What are its competitors?
ZOOPLANKTON AND PHYTOPLANKTON
BLADDER WRACK (AUTOTROPHE)
BARNACLE (HETEROTROPHE– FILTER FEEDER)
ENCRUSTING ALGA (AUTOTROPHE)
ANEMONE (HETEROTROPHE)
SHORE CRAB (HETEROTROPHE– OPPORTUNIST)
BLENNY (HETEROTROPHE– OPPORTUNIST)
A rock pool ecosystem is maintained by a series of interactions.
Abiotic factors (including temperature, volume of water, pH, salinity, dissolved oxygen substrate).
interacting with
The biotic community
N.B. The rock pool is not self-sustaining and relies on the twice-daily tidal cycle for maintenance of optimum abiotic (e.g. temperature, salinity) and biotic (e.g. nutrients, new colonizers) factors.
Autotrophes – the producers – which require an input of light and inorganic nutrients. Include algae and phytoplankton.
interacting with
Heterotrophes – the consumers – which include herbivores, carnivores, omnivores and decomposers

Energy flow through an ecosystem: I

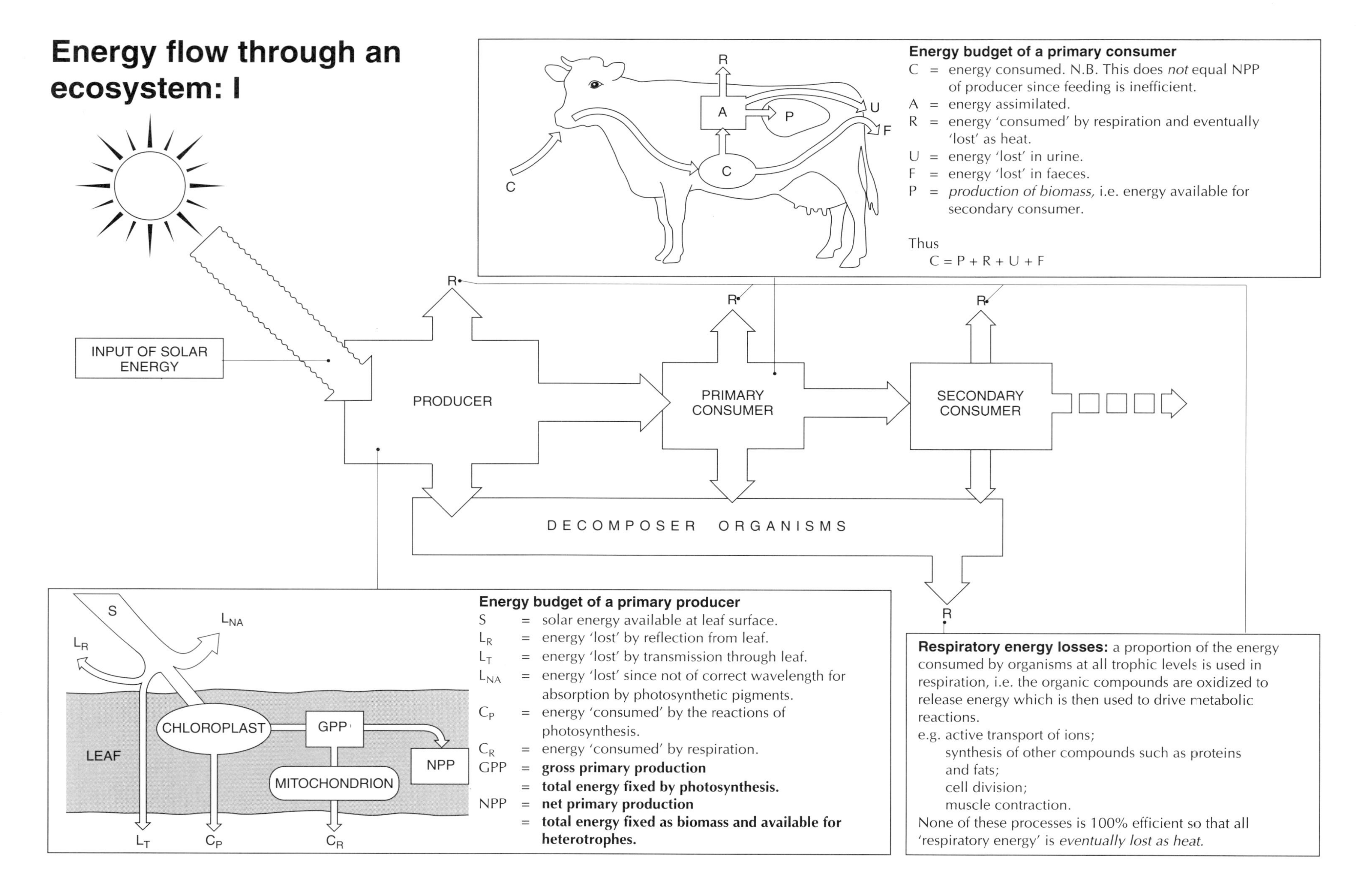

Energy budget of a primary consumer

C = energy consumed. N.B. This does *not* equal NPP of producer since feeding is inefficient.
A = energy assimilated.
R = energy 'consumed' by respiration and eventually 'lost' as heat.
U = energy 'lost' in urine.
F = energy 'lost' in faeces.
P = *production of biomass,* i.e. energy available for secondary consumer.

Thus
$C = P + R + U + F$

Energy budget of a primary producer

S = solar energy available at leaf surface.
L_R = energy 'lost' by reflection from leaf.
L_T = energy 'lost' by transmission through leaf.
L_{NA} = energy 'lost' since not of correct wavelength for absorption by photosynthetic pigments.
C_P = energy 'consumed' by the reactions of photosynthesis.
C_R = energy 'consumed' by respiration.
GPP = **gross primary production**
= **total energy fixed by photosynthesis.**
NPP = **net primary production**
= **total energy fixed as biomass and available for heterotrophes.**

Respiratory energy losses: a proportion of the energy consumed by organisms at all trophic levels is used in respiration, i.e. the organic compounds are oxidized to release energy which is then used to drive metabolic reactions.

e.g. active transport of ions;
synthesis of other compounds such as proteins and fats;
cell division;
muscle contraction.

None of these processes is 100% efficient so that all 'respiratory energy' is *eventually lost as heat.*

Energy flow through an ecosystem: II

A closed ecosystem is rare: migratory animals may deposit faeces, fruits and seeds may enter or leave during dispersal, and leaves may blow in from surrounding trees.

The *gross primary production (GPP)* – the total energy fixed by photosynthesis – represents only about 0.5–1% of the light energy available to the leaf.

At the equator, the *solar flux* (sunlight which reaches the Earth's upper atmosphere) is almost constant at 1.4 kJ m^{-2} s^{-1}. Most of this incoming sunlight energy is reflected by the atmosphere, heats the atmosphere and Earth's surface or causes the evaporation of water. Less than 0.1% actually falls on leaves and is thus available for photosynthesis.

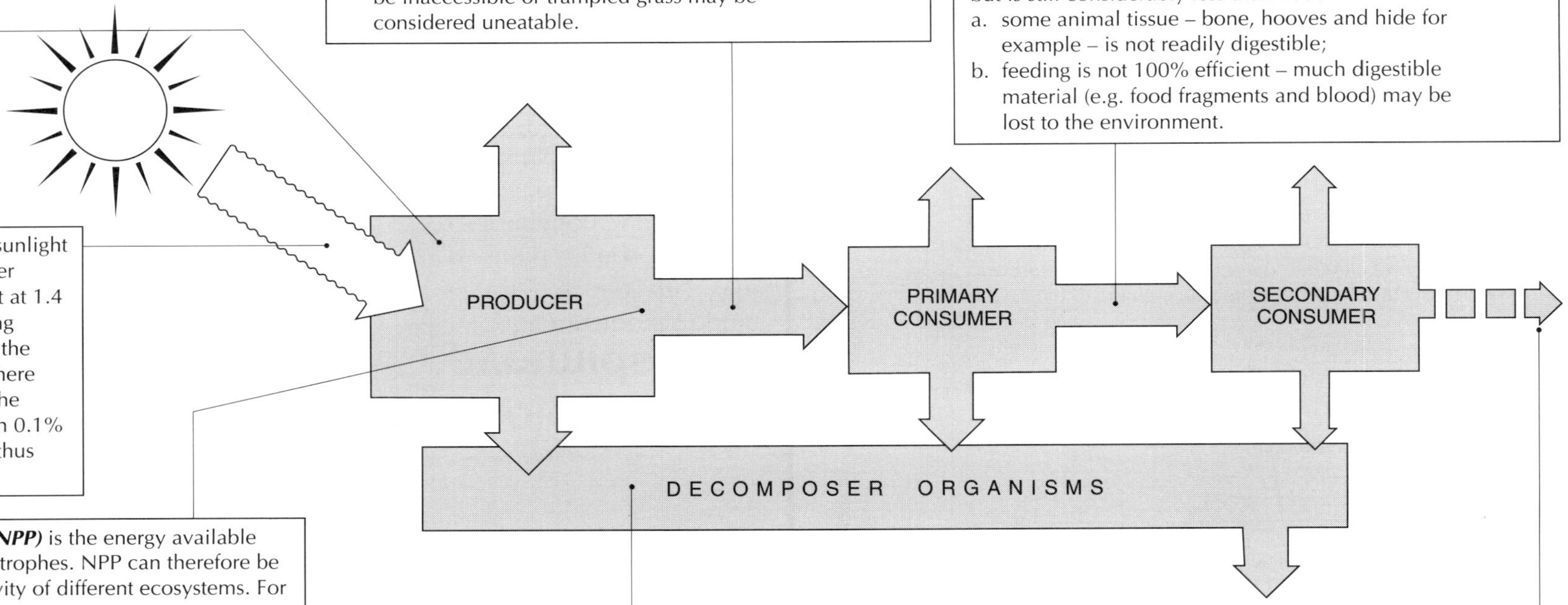

The ***net primary production (NPP)*** is the energy available for consumption by the heterotrophes. NPP can therefore be used to compare the productivity of different ecosystems. For example:

ECOSYSTEM	NPP (arbitrary units)
Coral reef	1000
Rainforest	880
Estuaries	600
Deciduous forest	500
Grassland	260
Open ocean	50
Desert	2

Productivity may be expressed as *units of energy* (e.g. kJ m^{-2} yr^{-1}) or *units of mass* (e.g. kg m^{-2} yr^{-1}).

Energy transfer from producer to primary consumer is typically in the order of 5–10% of NPP.
This is because
1. Much of plant biomass (NPP) is indigestible to herbivores – there are no animal enzymes to digest lignin and cellulose.
2. Much of the plant biomass may not be consumed by any individual herbivore species – roots may be inaccessible or trampled grass may be considered uneatable.

Energy transfer from primary consumer (herbivore) to secondary consumer (carnivore) is typically 10–20% of herbivore biomass. This is more efficient than producer → consumer because
1. animal tissue is more digestible than plant tissue;
2. animal tissue has a higher energy value;
3. carnivores may be extremely specialized for prey consumption;

but is still considerably less than 100% because
a. some animal tissue – bone, hooves and hide for example – is not readily digestible;
b. feeding is not 100% efficient – much digestible material (e.g. food fragments and blood) may be lost to the environment.

PRODUCER

PRIMARY CONSUMER

SECONDARY CONSUMER

DECOMPOSER ORGANISMS

The ***decomposers*** are fungi and bacteria which obtain energy and raw materials from animal and plant remains. In some situations 80% or more of the productivity at any trophic level may go through a decomposer pathway (e.g. forest floors of tropical forests). In some ecosystems – peat bogs, for example – the cold, wet, acidic conditions inhibit decomposition to such an extent that only about 10% of the material entering the decomposer food chain is broken down. The remainder accumulates as peat.

The limit to the number of trophic levels is determined by:
1. the total producer biomass;
2. the efficiency of energy transfer between trophic levels (only 10%).

In practice, the energy losses limit the number of levels to 3 or 4, very rarely 5 or 6. The longest food chains can only be supported by an enormous producer biomass, e.g. a 6 level chain will only have about 10% x 10% x 10% x 10% of NPP available to the top carnivores. The enormous volume of the oceans can provide sufficient biomass to support the longest food chains.

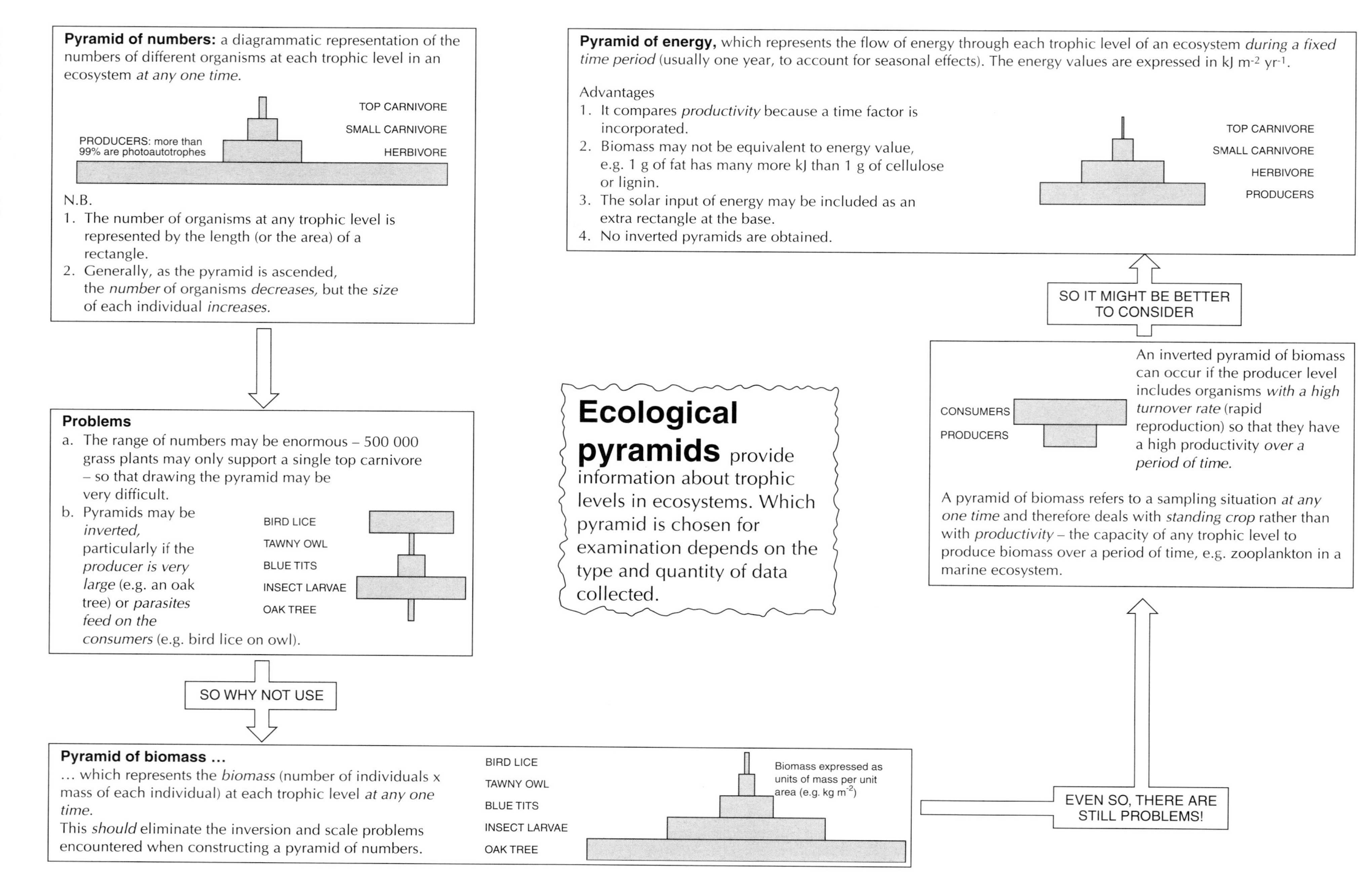
Ecological pyramids provide information about trophic levels in ecosystems. Which pyramid is chosen for examination depends on the type and quantity of data collected.
Pyramid of numbers: a diagrammatic representation of the numbers of different organisms at each trophic level in an ecosystem *at any one time.*
TOP CARNIVORE
SMALL CARNIVORE
HERBIVORE
PRODUCERS: more than 99% are photoautotrophes
N.B.
1. The number of organisms at any trophic level is represented by the length (or the area) of a rectangle.
2. Generally, as the pyramid is ascended, the *number* of organisms *decreases,* but the *size* of each individual *increases.*
Problems
a. The range of numbers may be enormous – 500 000 grass plants may only support a single top carnivore – so that drawing the pyramid may be very difficult.
b. Pyramids may be *inverted,* particularly if the *producer is very large* (e.g. an oak tree) or *parasites feed on the consumers* (e.g. bird lice on owl).
BIRD LICE
TAWNY OWL
BLUE TITS
INSECT LARVAE
OAK TREE
SO WHY NOT USE
Pyramid of biomass ...
... which represents the *biomass* (number of individuals x mass of each individual) at each trophic level *at any one time.*
This *should* eliminate the inversion and scale problems encountered when constructing a pyramid of numbers.
BIRD LICE
TAWNY OWL
BLUE TITS
INSECT LARVAE
OAK TREE
Biomass expressed as units of mass per unit area (e.g. kg m^{-2})
EVEN SO, THERE ARE STILL PROBLEMS!
CONSUMERS
PRODUCERS
An inverted pyramid of biomass can occur if the producer level includes organisms *with a high turnover rate* (rapid reproduction) so that they have a high productivity *over a period of time.*
A pyramid of biomass refers to a sampling situation *at any one time* and therefore deals with *standing crop* rather than with *productivity* – the capacity of any trophic level to produce biomass over a period of time, e.g. zooplankton in a marine ecosystem.
SO IT MIGHT BE BETTER TO CONSIDER
Pyramid of energy, which represents the flow of energy through each trophic level of an ecosystem *during a fixed time period* (usually one year, to account for seasonal effects). The energy values are expressed in kJ m^{-2} yr^{-1}.
Advantages
1. It compares *productivity* because a time factor is incorporated.
2. Biomass may not be equivalent to energy value, e.g. 1 g of fat has many more kJ than 1 g of cellulose or lignin.
3. The solar input of energy may be included as an extra rectangle at the base.
4. No inverted pyramids are obtained.
TOP CARNIVORE
SMALL CARNIVORE
HERBIVORE
PRODUCERS

Ecological succession proceeds via *several stages* to a *climax community* and is characterized by:

1. an increase in ***species diversity*** and in ***complexity of feeding relationships;***
2. a progressive increase in ***biomass;***
3. completion when ***energy input (community photosynthesis) = energy loss (community respiration).***

A COMMUNITY (all the species present in a given locality at any given time), is the group of interacting populations (all the members of a species in a place at a given time) which represents the biotic component of an ecosystem, and is seldom static. The relative abundance of different species may change, new species may enter the community and others may leave. There are reasons for these changes.

Catastrophes: may be natural (e.g. flooding, volcanic eruption) or caused by people (e.g. oil spill, deforestation).

Seasons: changes in temperature, rainfall, light intensity and windspeed, for example, may alter the suitability of a habitat for particular species.

Succession: long-term directional change in the composition of a community ***brought about by the actions of the organisms themselves.***

CAN BE EITHER

Primary succession occurs when the community develops on bare, uncolonized ground ***which has never had any vegetation growing on it***, e.g. mud in river deltas, lava flows, sand dunes, artificial ponds and newly erupted volcanic islands.

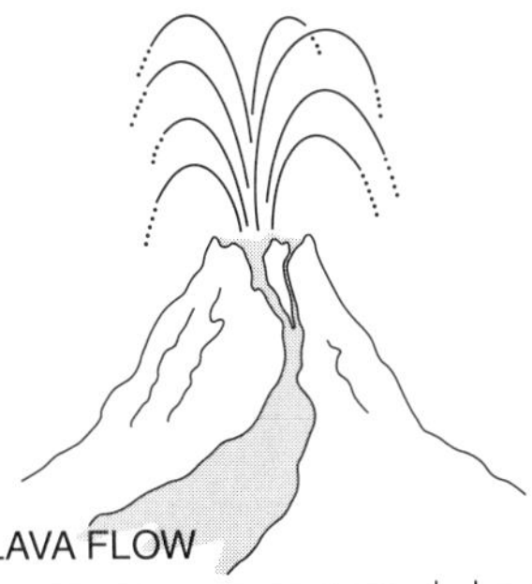

LAVA FLOW

OR

Secondary succession occurs on ground ***which had previously been colonized*** but is now available because the community has been destroyed, typically by fire, flood or as a result of human agricultural or industrial activities. Such ground will not be 'virgin' but will include remnants of soil, organic debris, seeds and even resistant animals and plants which have survived the changes, e.g. fire debris may be rich in minerals, particularly phosphate.

... BUT ALWAYS PROCEEDS VIA A SERIES OF STAGES

BARE GROUND ⟹ PIONEER COMMUNITY (COLONIZERS) ⟹ SECONDARY COMMUNITY ⟹ CLIMAX COMMUNITY

Migration: the arrival of seeds and spores. If conditions are suitable immigrant species may become established.

These species are simple plants, e.g. lichens and algae with minimal environmental demands. May show symbiotic relationships to aid their establishment. The community is ***open,*** i.e. space for further colonizers.

The number of species has risen – further stabilizing soil and adding nutrients. May still be an input of new species so that there will be both ***intra-*** and ***inter-specific competition.*** Pioneer species are often poor competitors and will be replaced by higher, more demanding plants such as grasses, shrubs and, eventually, trees.

The end point of succession: the community is now in equilibrium with the environment and is stable. Composition is often determined by one dominant species e.g. ***oak woodland.***

During succession each species modifies the environment, making it *more* suitable for new species and ***less*** suitable for those already there.

Each of these stages is called a ***seral stage*** and the complete succession is called a ***sere.***

Hydrosere: succession in an aquatic environment.
Xerosere: succession on dry land.
Halosere: succession in a salty environment.
Lithosere: succession on a rocky surface.

The carbon cycle depends upon both biochemical and physical processes.

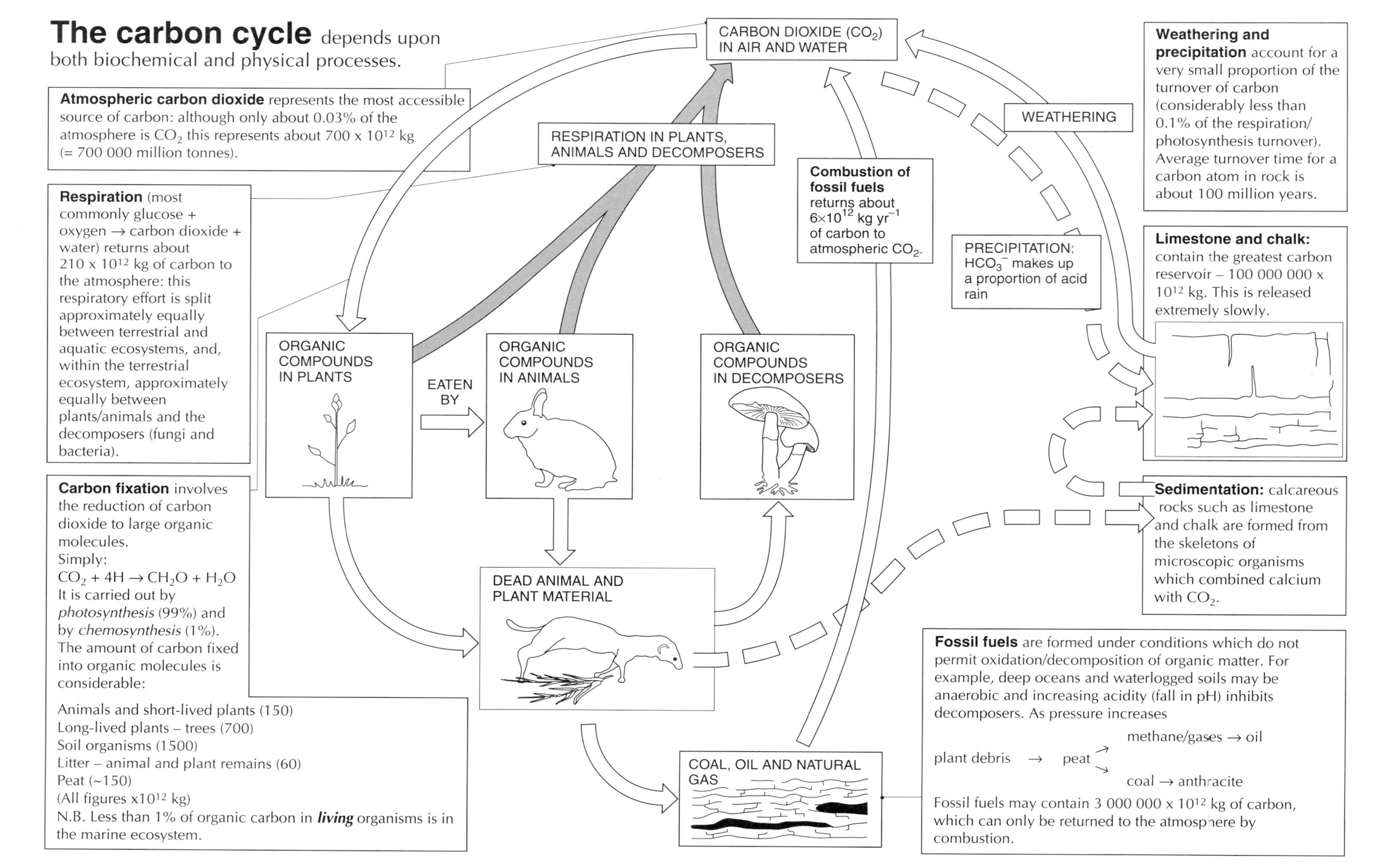

Atmospheric carbon dioxide represents the most accessible source of carbon: although only about 0.03% of the atmosphere is CO_2 this represents about 700×10^{12} kg (= 700 000 million tonnes).

Respiration (most commonly glucose + oxygen → carbon dioxide + water) returns about 210×10^{12} kg of carbon to the atmosphere: this respiratory effort is split approximately equally between terrestrial and aquatic ecosystems, and, within the terrestrial ecosystem, approximately equally between plants/animals and the decomposers (fungi and bacteria).

Carbon fixation involves the reduction of carbon dioxide to large organic molecules.
Simply:
$CO_2 + 4H \rightarrow CH_2O + H_2O$
It is carried out by *photosynthesis* (99%) and by *chemosynthesis* (1%).
The amount of carbon fixed into organic molecules is considerable:

Animals and short-lived plants (150)
Long-lived plants – trees (700)
Soil organisms (1500)
Litter – animal and plant remains (60)
Peat (~150)
(All figures $\times 10^{12}$ kg)
N.B. Less than 1% of organic carbon in ***living*** organisms is in the marine ecosystem.

Weathering and precipitation account for a very small proportion of the turnover of carbon (considerably less than 0.1% of the respiration/photosynthesis turnover). Average turnover time for a carbon atom in rock is about 100 million years.

Limestone and chalk: contain the greatest carbon reservoir – $100\,000\,000 \times 10^{12}$ kg. This is released extremely slowly.

Sedimentation: calcareous rocks such as limestone and chalk are formed from the skeletons of microscopic organisms which combined calcium with CO_2.

Fossil fuels are formed under conditions which do not permit oxidation/decomposition of organic matter. For example, deep oceans and waterlogged soils may be anaerobic and increasing acidity (fall in pH) inhibits decomposers. As pressure increases

plant debris → peat ↗ methane/gases → oil
plant debris → peat ↘ coal → anthracite

Fossil fuels may contain $3\,000\,000 \times 10^{12}$ kg of carbon, which can only be returned to the atmosphere by combustion.

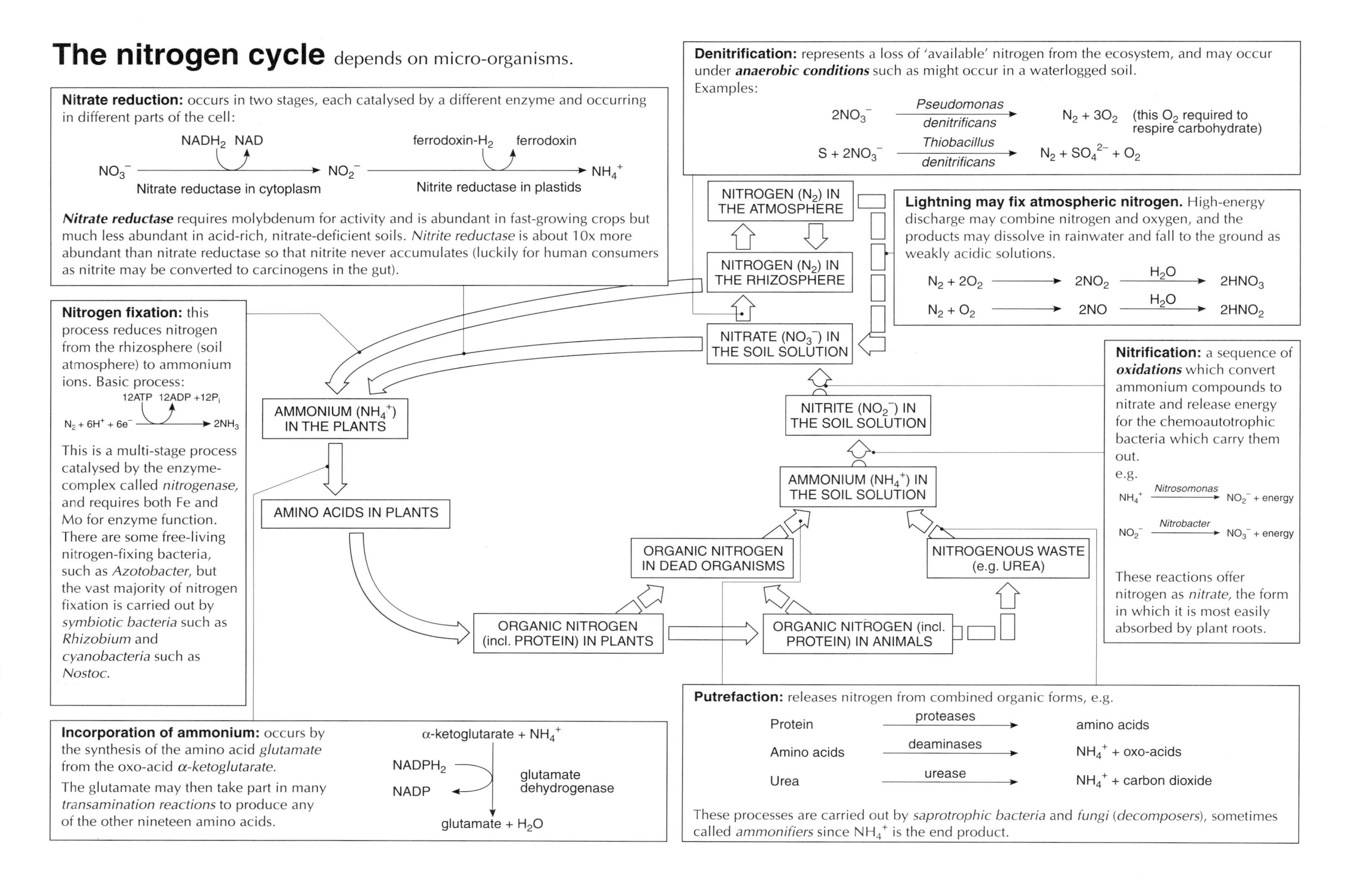

The nitrogen cycle depends on micro-organisms.
Nitrate reduction: occurs in two stages, each catalysed by a different enzyme and occurring in different parts of the cell:
NADH$_2$ NAD
ferrodoxin-H$_2$ ferrodoxin
NO_3^- → NO_2^- → NH_4^+
Nitrate reductase in cytoplasm
Nitrite reductase in plastids
Nitrate reductase requires molybdenum for activity and is abundant in fast-growing crops but much less abundant in acid-rich, nitrate-deficient soils. Nitrite reductase is about 10x more abundant than nitrate reductase so that nitrite never accumulates (luckily for human consumers as nitrite may be converted to carcinogens in the gut).
Denitrification: represents a loss of 'available' nitrogen from the ecosystem, and may occur under anaerobic conditions such as might occur in a waterlogged soil.
Examples:
$2NO_3^-$ → (Pseudomonas denitrificans) $N_2 + 3O_2$ (this O_2 required to respire carbohydrate)
$S + 2NO_3^-$ → (Thiobacillus denitrificans) $N_2 + SO_4^{2-} + O_2$
NITROGEN (N_2) IN THE ATMOSPHERE
NITROGEN (N_2) IN THE RHIZOSPHERE
NITRATE (NO_3^-) IN THE SOIL SOLUTION
Lightning may fix atmospheric nitrogen. High-energy discharge may combine nitrogen and oxygen, and the products may dissolve in rainwater and fall to the ground as weakly acidic solutions.
$N_2 + 2O_2$ → $2NO_2$ → (H_2O) $2HNO_3$
$N_2 + O_2$ → $2NO$ → (H_2O) $2HNO_2$
Nitrogen fixation: this process reduces nitrogen from the rhizosphere (soil atmosphere) to ammonium ions. Basic process:
12ATP 12ADP +12P$_i$
$N_2 + 6H^+ + 6e^-$ → $2NH_3$
This is a multi-stage process catalysed by the enzyme-complex called nitrogenase, and requires both Fe and Mo for enzyme function. There are some free-living nitrogen-fixing bacteria, such as Azotobacter, but the vast majority of nitrogen fixation is carried out by symbiotic bacteria such as Rhizobium and cyanobacteria such as Nostoc.
AMMONIUM (NH_4^+) IN THE PLANTS
AMINO ACIDS IN PLANTS
NITRITE (NO_2^-) IN THE SOIL SOLUTION
AMMONIUM (NH_4^+) IN THE SOIL SOLUTION
Nitrification: a sequence of oxidations which convert ammonium compounds to nitrate and release energy for the chemoautotrophic bacteria which carry them out.
e.g.
NH_4^+ → (Nitrosomonas) NO_2^- + energy
NO_2^- → (Nitrobacter) NO_3^- + energy
These reactions offer nitrogen as nitrate, the form in which it is most easily absorbed by plant roots.
ORGANIC NITROGEN IN DEAD ORGANISMS
NITROGENOUS WASTE (e.g. UREA)
ORGANIC NITROGEN (incl. PROTEIN) IN PLANTS
ORGANIC NITROGEN (incl. PROTEIN) IN ANIMALS
Incorporation of ammonium: occurs by the synthesis of the amino acid glutamate from the oxo-acid α-ketoglutarate.
The glutamate may then take part in many transamination reactions to produce any of the other nineteen amino acids.
α-ketoglutarate + NH_4^+
NADPH$_2$
NADP
glutamate dehydrogenase
glutamate + H_2O
Putrefaction: releases nitrogen from combined organic forms, e.g.
Protein → (proteases) amino acids
Amino acids → (deaminases) NH_4^+ + oxo-acids
Urea → (urease) NH_4^+ + carbon dioxide
These processes are carried out by saprotrophic bacteria and fungi (decomposers), sometimes called ammonifiers since NH_4^+ is the end product.

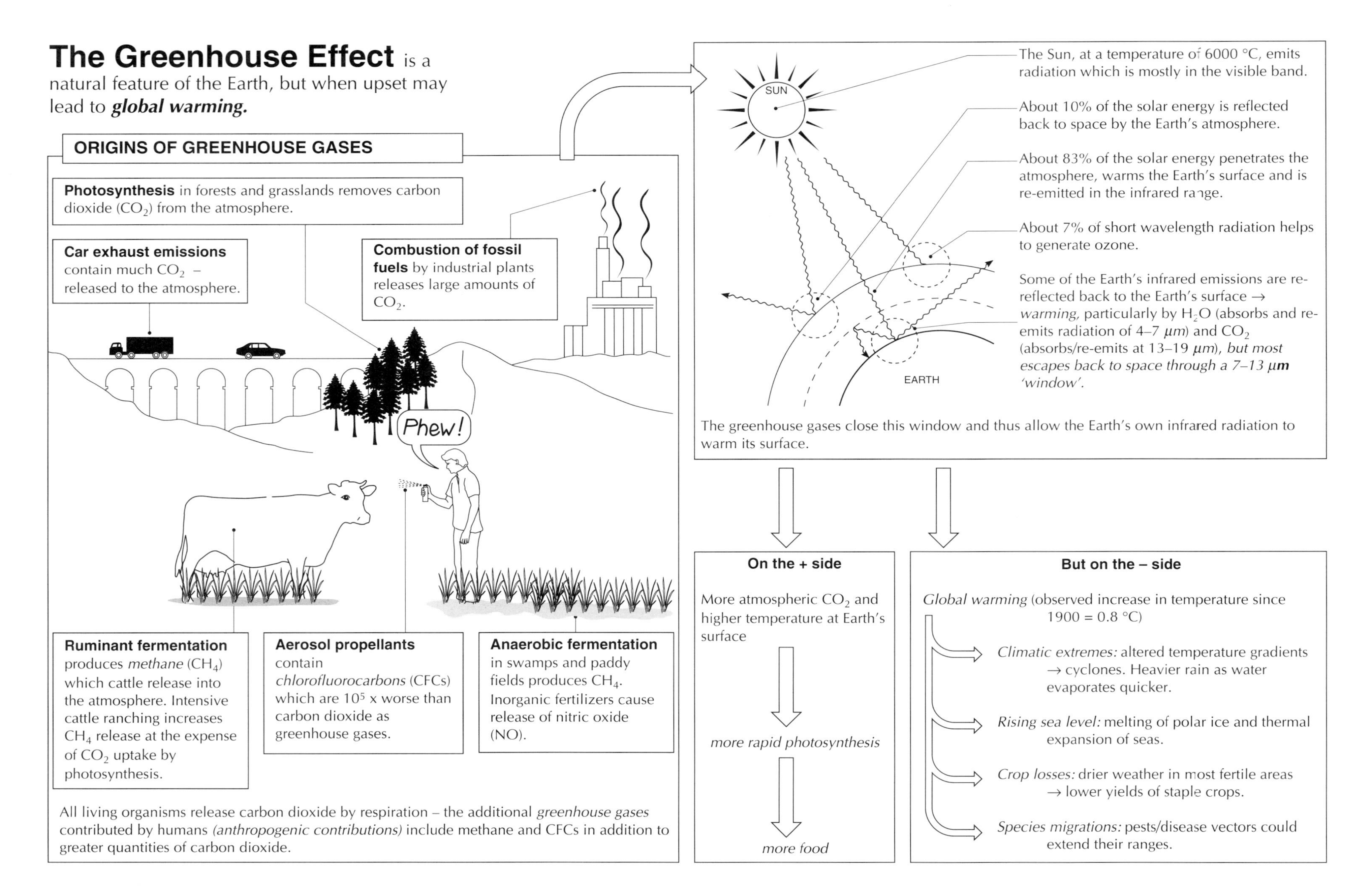
The Greenhouse Effect is a natural feature of the Earth, but when upset may lead to global warming.
ORIGINS OF GREENHOUSE GASES
Photosynthesis in forests and grasslands removes carbon dioxide (CO_2) from the atmosphere.
Car exhaust emissions contain much CO_2 – released to the atmosphere.
Combustion of fossil fuels by industrial plants releases large amounts of CO_2.
Phew!
Ruminant fermentation produces methane (CH_4) which cattle release into the atmosphere. Intensive cattle ranching increases CH_4 release at the expense of CO_2 uptake by photosynthesis.
Aerosol propellants contain chlorofluorocarbons (CFCs) which are 10^5 x worse than carbon dioxide as greenhouse gases.
Anaerobic fermentation in swamps and paddy fields produces CH_4. Inorganic fertilizers cause release of nitric oxide (NO).
All living organisms release carbon dioxide by respiration – the additional greenhouse gases contributed by humans (anthropogenic contributions) include methane and CFCs in addition to greater quantities of carbon dioxide.
SUN
EARTH
The Sun, at a temperature of 6000 °C, emits radiation which is mostly in the visible band.
About 10% of the solar energy is reflected back to space by the Earth's atmosphere.
About 83% of the solar energy penetrates the atmosphere, warms the Earth's surface and is re-emitted in the infrared range.
About 7% of short wavelength radiation helps to generate ozone.
Some of the Earth's infrared emissions are re-reflected back to the Earth's surface → warming, particularly by H_2O (absorbs and re-emits radiation of 4–7 μm) and CO_2 (absorbs/re-emits at 13–19 μm), but most escapes back to space through a 7–13 μm 'window'.
The greenhouse gases close this window and thus allow the Earth's own infrared radiation to warm its surface.
On the + side
More atmospheric CO_2 and higher temperature at Earth's surface
more rapid photosynthesis
more food
But on the – side
Global warming (observed increase in temperature since 1900 = 0.8 °C)
Climatic extremes: altered temperature gradients → cyclones. Heavier rain as water evaporates quicker.
Rising sea level: melting of polar ice and thermal expansion of seas.
Crop losses: drier weather in most fertile areas → lower yields of staple crops.
Species migrations: pests/disease vectors could extend their ranges.

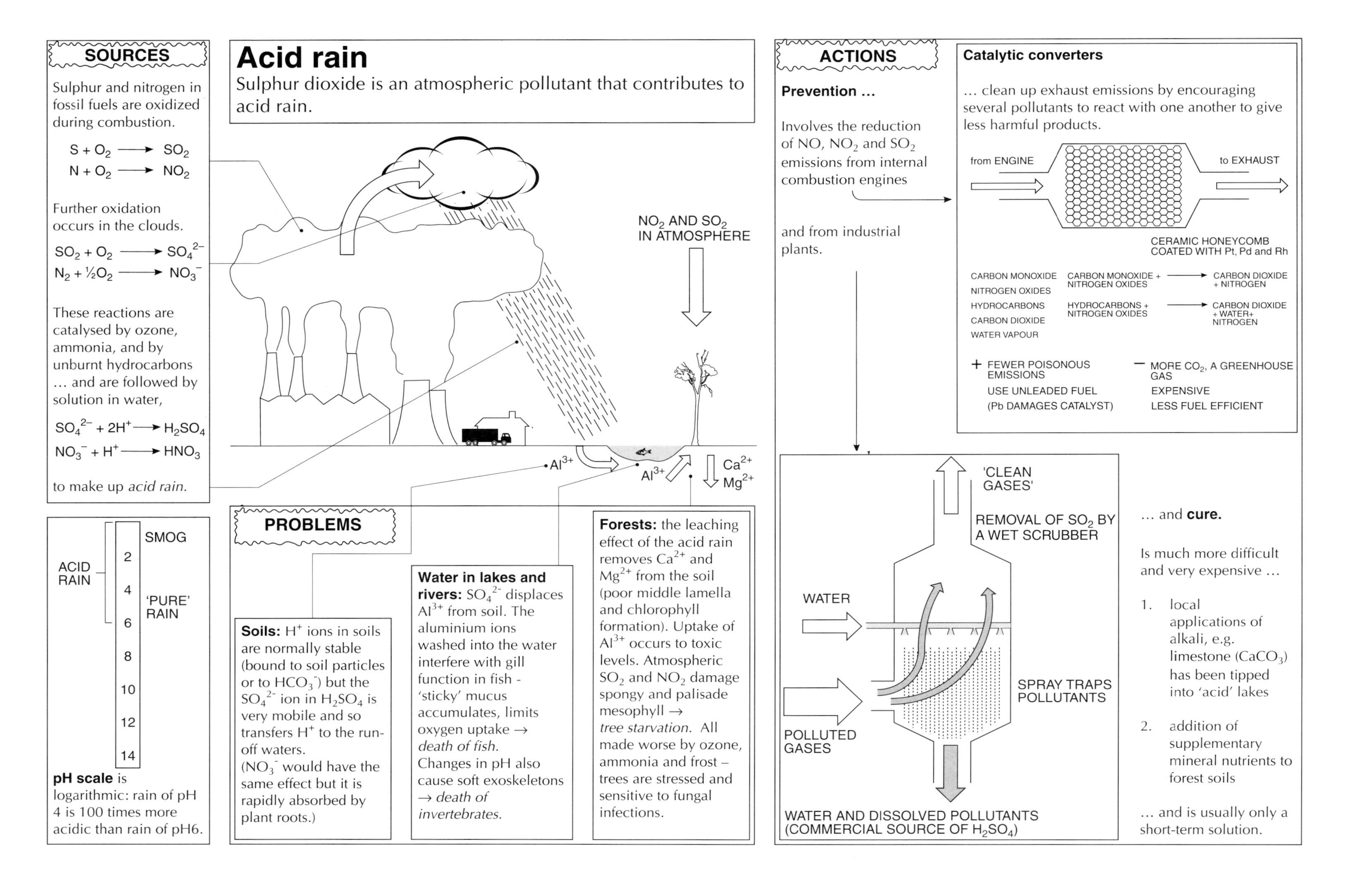
SOURCES
Sulphur and nitrogen in fossil fuels are oxidized during combustion.
$S + O_2 \longrightarrow SO_2$
$N + O_2 \longrightarrow NO_2$
Further oxidation occurs in the clouds.
$SO_2 + O_2 \longrightarrow SO_4^{2-}$
$N_2 + \frac{1}{2}O_2 \longrightarrow NO_3^-$
These reactions are catalysed by ozone, ammonia, and by unburnt hydrocarbons ... and are followed by solution in water,
$SO_4^{2-} + 2H^+ \longrightarrow H_2SO_4$
$NO_3^- + H^+ \longrightarrow HNO_3$
to make up *acid rain.*
Acid rain
Sulphur dioxide is an atmospheric pollutant that contributes to acid rain.
NO_2 AND SO_2 IN ATMOSPHERE
Al^{3+}
Al^{3+}
Ca^{2+}
Mg^{2+}
ACID RAIN
SMOG
'PURE' RAIN
2
4
6
8
10
12
14
pH scale is logarithmic: rain of pH 4 is 100 times more acidic than rain of pH6.
PROBLEMS
Soils: H^+ ions in soils are normally stable (bound to soil particles or to HCO_3^-) but the SO_4^{2-} ion in H_2SO_4 is very mobile and so transfers H^+ to the run-off waters. (NO_3^- would have the same effect but it is rapidly absorbed by plant roots.)
Water in lakes and rivers: SO_4^{2-} displaces Al^{3+} from soil. The aluminium ions washed into the water interfere with gill function in fish - 'sticky' mucus accumulates, limits oxygen uptake → *death of fish.* Changes in pH also cause soft exoskeletons → *death of invertebrates.*
Forests: the leaching effect of the acid rain removes Ca^{2+} and Mg^{2+} from the soil (poor middle lamella and chlorophyll formation). Uptake of Al^{3+} occurs to toxic levels. Atmospheric SO_2 and NO_2 damage spongy and palisade mesophyll → *tree starvation.* All made worse by ozone, ammonia and frost – trees are stressed and sensitive to fungal infections.
ACTIONS
Prevention ...
Involves the reduction of NO, NO_2 and SO_2 emissions from internal combustion engines
and from industrial plants.
Catalytic converters
... clean up exhaust emissions by encouraging several pollutants to react with one another to give less harmful products.
from ENGINE
to EXHAUST
CERAMIC HONEYCOMB COATED WITH Pt, Pd and Rh
CARBON MONOXIDE
NITROGEN OXIDES
HYDROCARBONS
CARBON DIOXIDE
WATER VAPOUR
CARBON MONOXIDE + NITROGEN OXIDES → CARBON DIOXIDE + NITROGEN
HYDROCARBONS + NITROGEN OXIDES → CARBON DIOXIDE + WATER+ NITROGEN
+ FEWER POISONOUS EMISSIONS
USE UNLEADED FUEL
(Pb DAMAGES CATALYST)
– MORE CO_2, A GREENHOUSE GAS
EXPENSIVE
LESS FUEL EFFICIENT
'CLEAN GASES'
REMOVAL OF SO_2 BY A WET SCRUBBER
WATER
SPRAY TRAPS POLLUTANTS
POLLUTED GASES
WATER AND DISSOLVED POLLUTANTS (COMMERCIAL SOURCE OF H_2SO_4)
... and **cure.**
Is much more difficult and very expensive ...
1. local applications of alkali, e.g. limestone ($CaCO_3$) has been tipped into 'acid' lakes
2. addition of supplementary mineral nutrients to forest soils
... and is usually only a short-term solution.

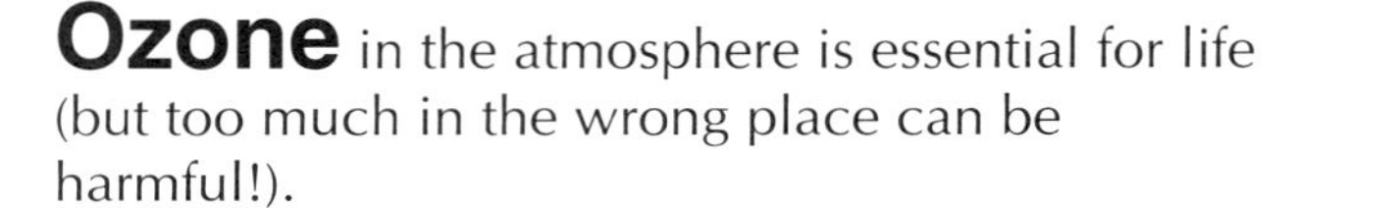

Ozone in the atmosphere is essential for life (but too much in the wrong place can be harmful!).

'Holes' in the ozone layer: measurements made by British scientists at Halley Bay, Antarctica, showed a thinning of the ozone layer caused by an accumulation of atmospheric chlorine during the winter months. The effect is partially reversed in the summer and may be peculiar to the Antarctic, but is seen as a warning that we must reduce production of long-lived CFCs.

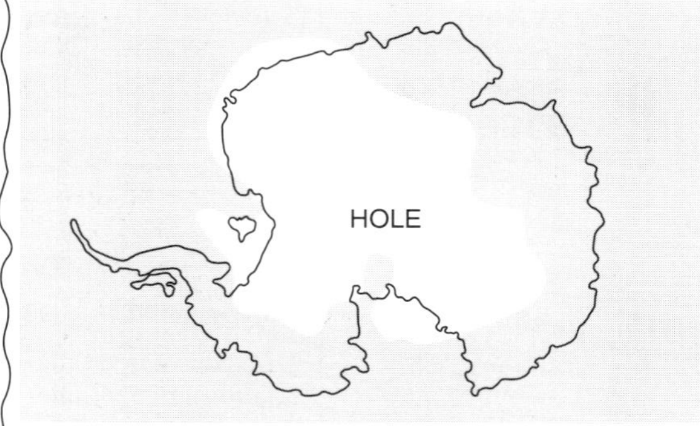

Antarctica: ozone layer thinned over an area as large as the United States.

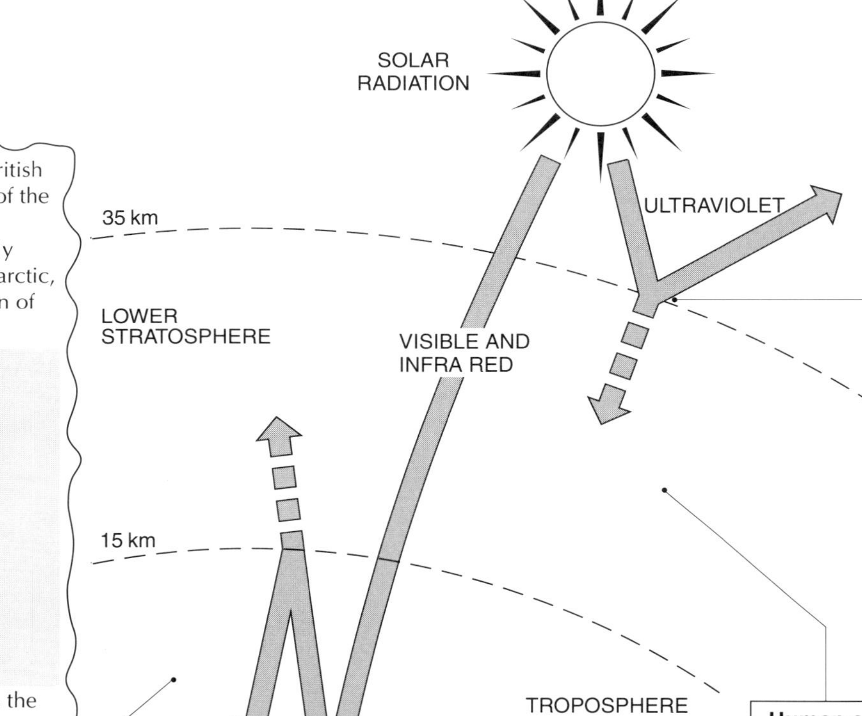

Tropospheric (low level) ozone causes problems

1. It acts as a *greenhouse gas,* absorbing and re-radiating heat which raises the temperature at the Earth's surface.
2. As a result of 1, *thermal inversion* occurs – a layer of warm air traps cool air (containing dust and smoke) close to the Earth causing *smog.*
3. It causes *irritation* of eyes, throat and lungs and may cause death in sufferers from respiratory ailments as breathing is impaired.
4. It severely damages the photosynthetic mesophyll layers of plants by forming powerful oxidizing free radicals: this may lead to a *10% reduction in crop production.*

Human activities generate ozone in the troposphere

Fossil fuel combustion ⟶ NO_2

NO_2 —sunlight *⟶ $NO + O$

$O + O_2$ ⟶ O_3 : OZONE

* This occurs more rapidly in the presence of unburned hydrocarbons.

Stratospheric (high level) ozone offers protection

Ozone absorbs *solar ultraviolet radiation* which would otherwise reach the Earth's surface.

UV-B (290–320 nm) already reaches the Earth

⟶ sunburn
some forms of skin cancer
cataract

One estimate suggests that a 2.5% reduction in the ozone layer would cause 0.8 million cancer deaths and 40 000 000 additional cases of skin cancer

⟶ reduced productivity of some plant species, e.g. soya bean.

UV-C (240–290 nm) does not at present reach the Earth but in laboratory tests has been shown to increase damage to DNA (more mutations) and to proteins.

Human activities reduce ozone in the stratosphere

CFCs (chlorofluorocarbons) used in:
refrigerator coolants;
aerosols;
expanded plastics.

Atmospheric chlorine

$Cl + O_3$ ⟶ $ClO + O_2$

$ClO + O$ ⟶ $Cl + O_2$

i.e. *ozone levels are depleted.*
Chlorine is released to degrade further ozone molecules.

Deforestation: the rapid destruction of woodland.

Has been occurring on a major scale throughout the world.

- Between 1880 and1980 about 40% of all tropical rainforest was destroyed.
- Britain has fallen from 85% forest cover to about 8% (probably the lowest in Europe).
- Major reasons
  - -removal of hardwood for high-quality furnishings;
  - -removal of softwoods for chipboards, paper and other wood products;
  - -clearance for cattle ranching and for cash crop agriculture;
  - -clearance for urban development (roads and towns being built).

Current losses: about 11 hectares ***per minute*** (that's about 40 soccer or hockey pitches!).

Reduction in soil fertility

1. Deciduous trees may contain 90% of the nutrients in a forest ecosystem: these nutrients are removed, and are thus not available to the soil, if the trees are cut down and taken away.
2. Soil erosion may be rapid since in the absence of trees
   a. wind and direct rain may remove the soil;
   b. soil structure is no longer stabilized by tree root systems.

N.B. The soil below coniferous forests is often of poor quality for agriculture because the shed pine needles contain toxic compounds which act as germination and growth inhibitors.

Flooding and landslips
Heavy rainfall on deforested land is not 'held up': normally 25% of rainfall is absorbed by foliage or evaporates and 50% is absorbed by root systems. As a result water may accumulate rapidly in river valleys, often causing landslips from steep hillsides.

Changes in recycling of materials: Fewer trees mean

1. atmospheric CO_2 concentration may rise as less CO_2 is removed for photosynthesis;
2. atmospheric O_2 – vital for aerobic respiration – is diminished as less is produced by photosynthesis;
3. the atmosphere may become drier and the soil wetter as evaporation (from soil) is slower than transpiration (from trees).

Climatic changes

1. Reduced transpiration rates and drier atmosphere affect the water cycle and reduce rainfall.
2. Rapid heat absorption by bare soil raises the temperature of the lower atmosphere in some areas, causing thermal gradients which result in more frequent and intense winds.

Species extinction: Many species are dependent on forest conditions.

e.g. mountain gorilla depends on cloud forest of Central Africa; golden lion tamarin depends on coastal rainforest of Brazil; osprey depends on mature pine forests in Northern Europe.

It is estimated that one plant and one animal species become extinct every 30 minutes due to deforestation.

Many plant species may have medicinal properties,
e.g. as tranquilizers, reproductive hormones, anticoagulants, painkillers and antibiotics.
The Madagascan periwinkle, for example, yields one of the most potent known anti-leukaemia drugs.

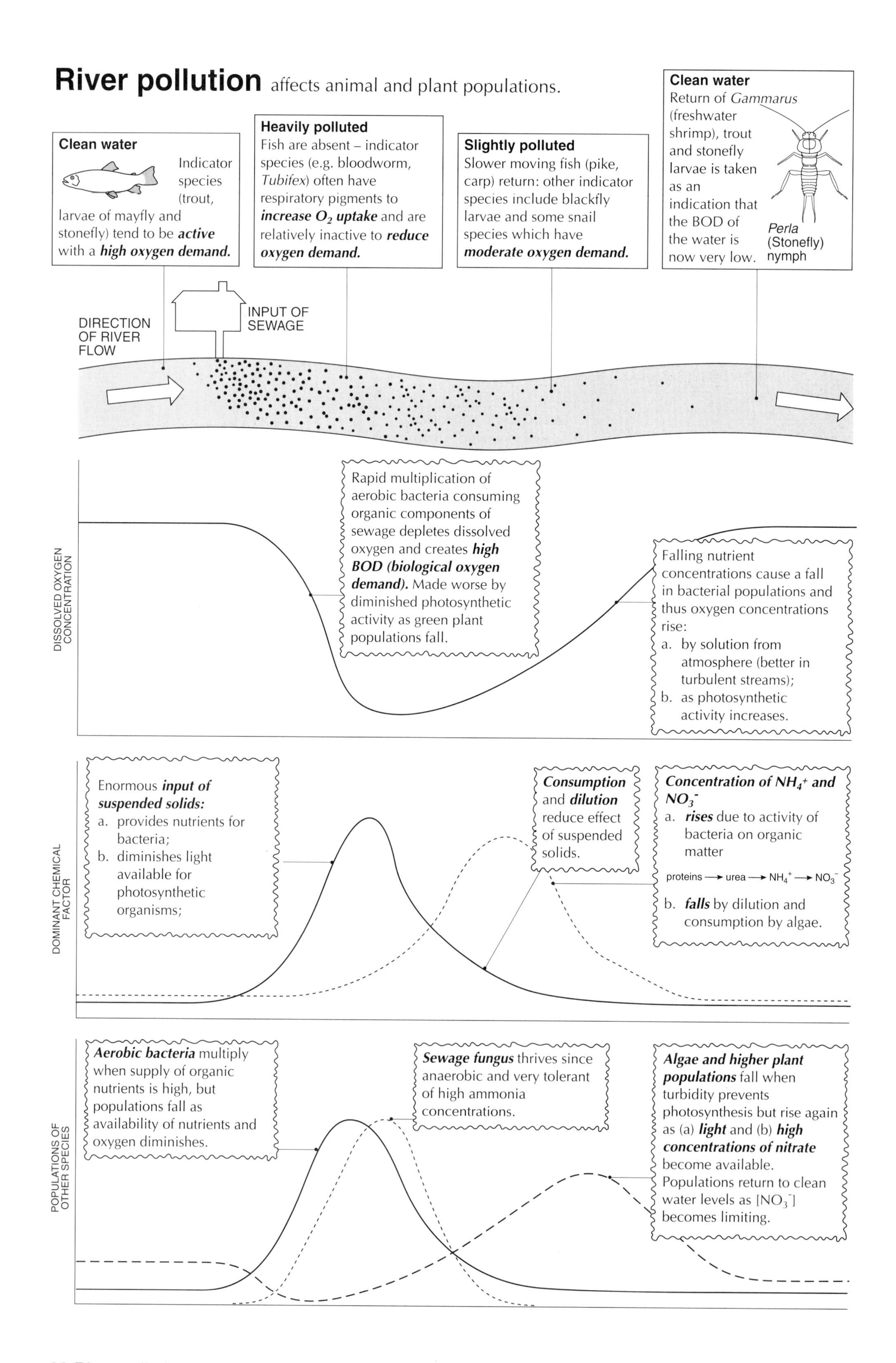
River pollution affects animal and plant populations.
Clean water
Indicator species (trout, larvae of mayfly and stonefly) tend to be *active* with a *high oxygen demand.*
Heavily polluted
Fish are absent – indicator species (e.g. bloodworm, *Tubifex*) often have respiratory pigments to *increase O_2 uptake* and are relatively inactive to *reduce oxygen demand.*
Slightly polluted
Slower moving fish (pike, carp) return: other indicator species include blackfly larvae and some snail species which have *moderate oxygen demand.*
Clean water
Return of *Gammarus* (freshwater shrimp), trout and stonefly larvae is taken as an indication that the BOD of the water is now very low.
Perla (Stonefly) nymph
DIRECTION OF RIVER FLOW
INPUT OF SEWAGE
DISSOLVED OXYGEN CONCENTRATION
Rapid multiplication of aerobic bacteria consuming organic components of sewage depletes dissolved oxygen and creates *high BOD (biological oxygen demand).* Made worse by diminished photosynthetic activity as green plant populations fall.
Falling nutrient concentrations cause a fall in bacterial populations and thus oxygen concentrations rise:
a. by solution from atmosphere (better in turbulent streams);
b. as photosynthetic activity increases.
DOMINANT CHEMICAL FACTOR
Enormous *input of suspended solids:*
a. provides nutrients for bacteria;
b. diminishes light available for photosynthetic organisms;
Consumption and *dilution* reduce effect of suspended solids.
Concentration of NH_4^+ and NO_3^-
a. *rises* due to activity of bacteria on organic matter
proteins → urea → NH_4^+ → NO_3^-
b. *falls* by dilution and consumption by algae.
POPULATIONS OF OTHER SPECIES
Aerobic bacteria multiply when supply of organic nutrients is high, but populations fall as availability of nutrients and oxygen diminishes.
Sewage fungus thrives since anaerobic and very tolerant of high ammonia concentrations.
Algae and higher plant populations fall when turbidity prevents photosynthesis but rise again as (a) *light* and (b) *high concentrations of nitrate* become available. Populations return to clean water levels as $[NO_3^-]$ becomes limiting.

Nitrates are significant pollutants of water.

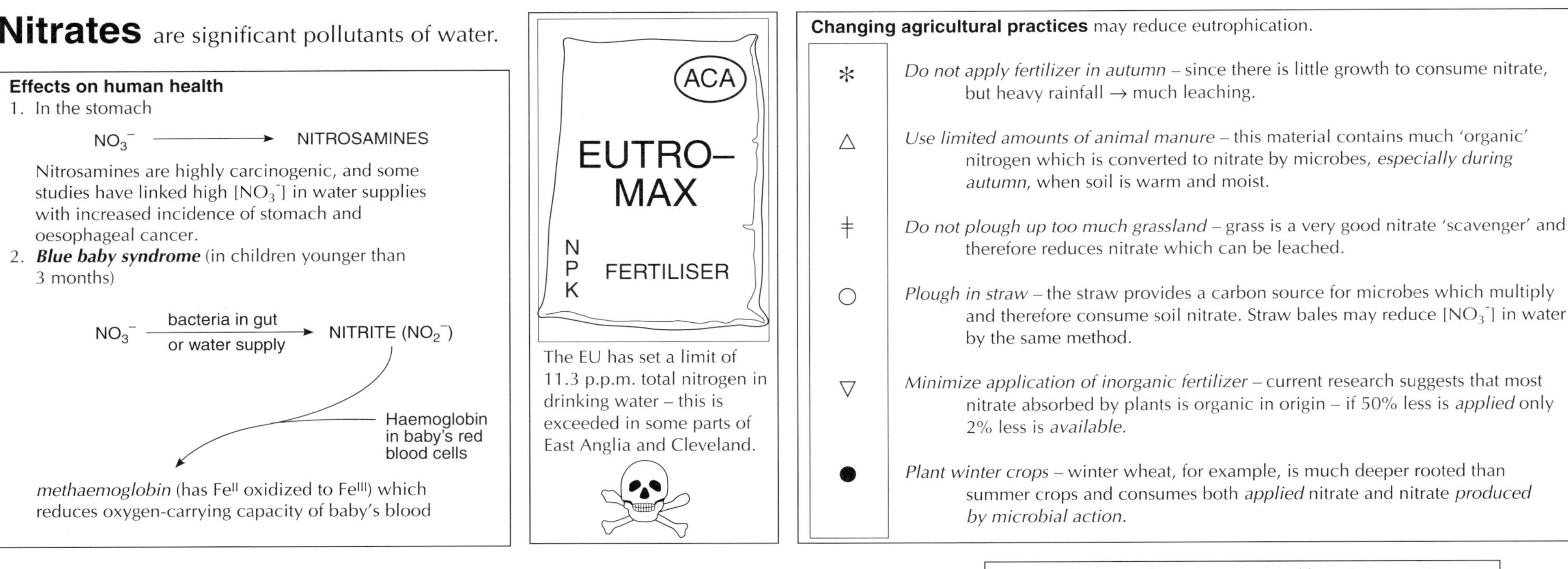

Effects on human health

1. In the stomach

NO_3^- ⟶ NITROSAMINES

Nitrosamines are highly carcinogenic, and some studies have linked high $[NO_3^-]$ in water supplies with increased incidence of stomach and oesophageal cancer.

2. ***Blue baby syndrome*** (in children younger than 3 months)

NO_3^- —bacteria in gut or water supply→ NITRITE (NO_2^-)

methaemoglobin (has Fe^{II} oxidized to Fe^{III}) which reduces oxygen-carrying capacity of baby's blood

The EU has set a limit of 11.3 p.p.m. total nitrogen in drinking water – this is exceeded in some parts of East Anglia and Cleveland.

Changing agricultural practices may reduce eutrophication.

✻ *Do not apply fertilizer in autumn* – since there is little growth to consume nitrate, but heavy rainfall → much leaching.

△ *Use limited amounts of animal manure* – this material contains much 'organic' nitrogen which is converted to nitrate by microbes, *especially during autumn*, when soil is warm and moist.

ǂ *Do not plough up too much grassland* – grass is a very good nitrate 'scavenger' and therefore reduces nitrate which can be leached.

○ *Plough in straw* – the straw provides a carbon source for microbes which multiply and therefore consume soil nitrate. Straw bales may reduce $[NO_3^-]$ in water by the same method.

▽ *Minimize application of inorganic fertilizer* – current research suggests that most nitrate absorbed by plants is organic in origin – if 50% less is *applied* only 2% less is *available*.

● *Plant winter crops* – winter wheat, for example, is much deeper rooted than summer crops and consumes both *applied* nitrate and nitrate *produced by microbial action*.

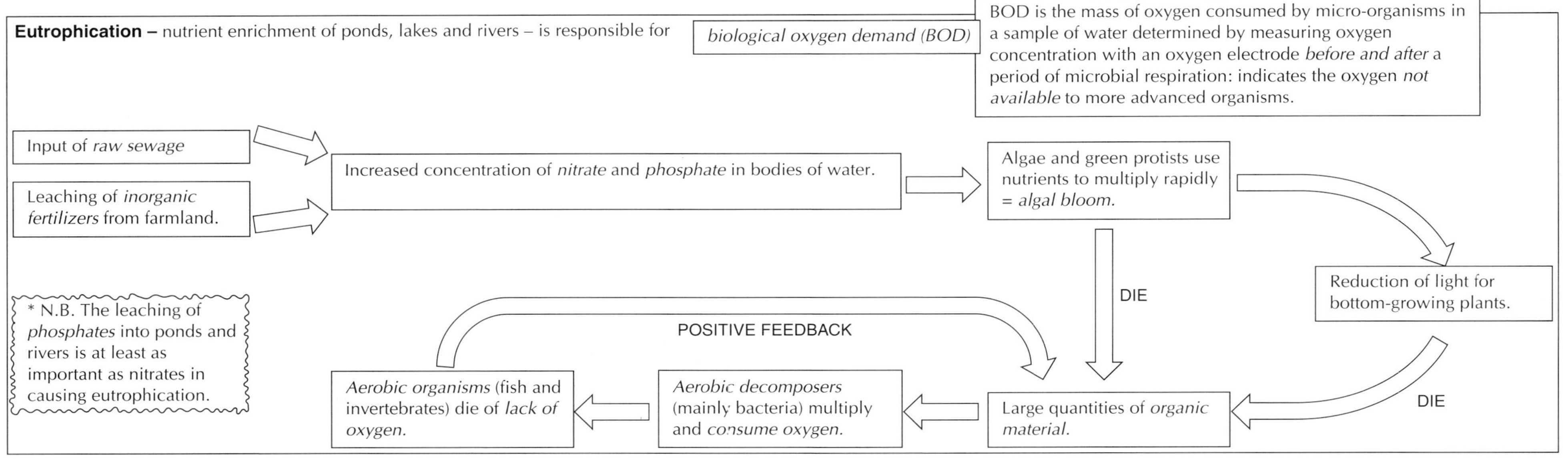

Eutrophication – nutrient enrichment of ponds, lakes and rivers – is responsible for *biological oxygen demand (BOD)*

BOD is the mass of oxygen consumed by micro-organisms in a sample of water determined by measuring oxygen concentration with an oxygen electrode *before and after* a period of microbial respiration: indicates the oxygen *not available* to more advanced organisms.

* N.B. The leaching of *phosphates* into ponds and rivers is at least as important as nitrates in causing eutrophication.

Chemical pest control may involve the use of:

herbicides – for control of weeds;

insecticides – for control of insects;

fungicides – for control of fungi;

molluscicides – for control of slugs and snails.

PROBLEMS WITH INSECTICIDES: these arise since the principal idea behind chemical control is to *kill as many of the pests as possible* – the effects on harmless or beneficial organisms were not studied or were ignored.

1. **Direct killing:** accidental misuse of toxic chemicals may cause death in humans or in domestic animals.
2. **Non-specificity:** non-target species, particularly natural predators of the pest species, may be killed by some wide-spectrum insecticides, e.g. large doses of ***dieldrin*** killed many birds as well as the Japanese beetle pest which was the intended target organism.
3. **Pest resistance:** genetic variation means that each pest population contains a *few* resistant individuals. The pesticide eliminates the non-resistant forms and thus a resistant population is selected for and may quickly develop (since many pests reproduce rapidly).
4. **Pest replacement:** most crops are susceptible to attack by more than one species – a ***pest complex*** and the use of a pesticide to eliminate one species may simply allow another species to assume major pest proportions (since a pesticide may be more deadly to one species than another).
5. **Pest resurgence:** non-specific pesticides may kill natural predators as well as pests – a small residual pest population may now multiply without check, creating a worse problem than initially was present.
6. **Bioaccumulation of toxins:** pesticides or their products may be toxic
   a. they may seriously affect micro-organisms and thus alter decomposition in soils;
   b. they may pass along food chains, becoming more concentrated in organisms further up the chain.

e.g. DDT used as an insecticide accumulates in the fatty tissues of carnivorous animals, inhibiting cytochrome oxidase and limiting reproductive success (especially thin eggshells in birds of prey).

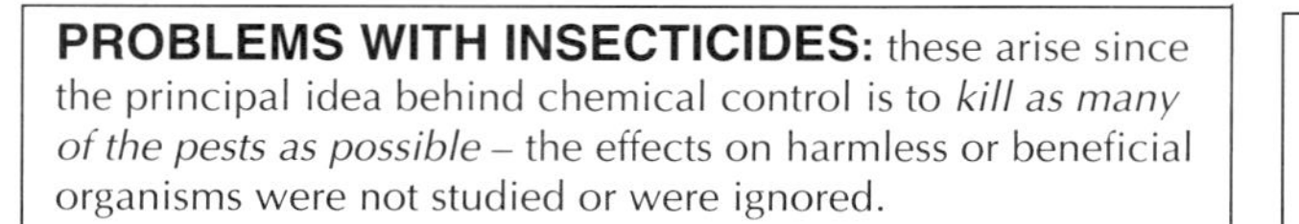

Relative DDT concentration along an aquatic food chain.

HERBICIDES may be

1. **Pre-emergent,** i.e. applied *before* emergence of crop.
   a. Contact herbicides, e.g. *Paraquat,* which kill all above-ground parts of all plants.
   b. Residual herbicides, e.g. *Linuron,* which bind to soil particles and kill weed seedlings as they emerge.

Pre-emergent herbicides can be *non-selective* and are ideal for clearing ground prior to cultivation.

2. **Post-emergent** is applied to both crop and weed, and therefore must be *selective.* Many, such as 2,4-D, are growth regulators.

Systemic herbicides, such as *glyphosate,* are absorbed by weeds and translocated to the meristems where they typically act by inhibition of cell division.

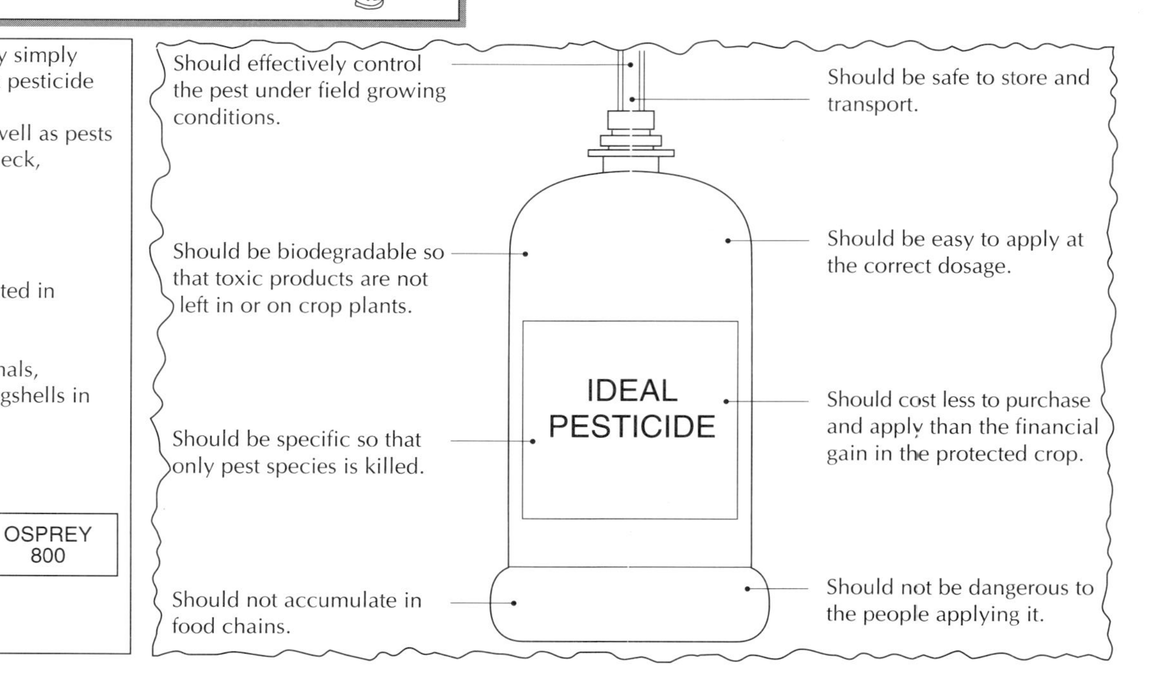

Biological pest control reduces the population of one species to levels at which it is no longer a pest by the use of one of the pest species' natural predators.

IDEAL RELATIONSHIP BETWEEN PEST AND ITS CONTROL AGENT

The *pest species* becomes the *prey* of the control agent: it is the *target* in the system of biological control.

Pest population falls due to *predation* by control agent.

Population size above which the pest is *economically harmful:* often determined by the expected yield and potential value of the crop.

Population of control agent falls because of a food shortage caused by reduction in prey (pest) numbers.

SIZE OF POPULATION / arbitrary units

TIME / arbitrary units

Control agent selected according to the criteria outlined above is the *predator* on the prey *pest* species. Population rises as the agent breeds, if conditions are appropriate.

Introduction of control agent: the size of the introduced population must be great enough to ensure a rise in numbers which is rapid enough to ensure control of the pest population within an economic time, e.g. before a crop plant has been extensively damaged.

A *dynamic equilibrium* is set up in which a moderate residual population of the control agent is able to permanently restrict the population of the pest. N.B. the pest species *must not be entirely eliminated* or the control agent will die out and a further introduction will be necessary to prevent re-establishment of economically damaging pest populations.

TYPICAL BIOLOGICAL CONTROL PROGRAMME

1. Identify the pest and *trace its origins,* i.e. where did it come from?
2. Investigate original site of pest and *identify natural enemies* of the pest.
3. Test the potential *control agent* under careful quarantine to ensure
   a. that it is *specific* (does not prey on other species)
   b. will not change its prey species and *become a pest itself*
   c. has a life cycle which will allow it to *develop a population large enough to act as an economic control.*
4. *Mass culture* of the control agent.
5. Development of the most *effective distribution/release method* for the control agent.

PRINCIPAL TECHNIQUES IN BIOLOGICAL CONTROL

1. Use a *herbivore* to control a *weed* species,
   e.g. *Cactoblastis* larvae on prickly pear.
2. Use a *carnivore* to control a *herbivorous pest,*
   e.g. hoverfly larvae on aphids.
3. Use a *parasite* to control its *host,*
   e.g. *Encharsia,* a parasitic wasp, on the greenhouse whitefly, *Trialeurodes vaporariorum.*
4. Disrupt the *breeding cycle* of a pest *if it mates once only in its life,*
   e.g. release of sterilized males of the screw worm fly, a flesh eating parasite of cattle.

5. Control of *pest behaviour,*
   e.g. sex attractant pheromones are used to attract apple codling moths into lethal traps.

An ideal human diet

An ideal human diet contains fat, protein, carbohydrate, vitamins, minerals, water and fibre ***in the correct proportions.***

Carbohydrates

Principally as a *respiratory substrate,* i.e. to be oxidized to release *energy* for active transport, synthesis of macromolecules, cell division and muscle contraction.

Common sources: rice, potatoes, wheat and other cereal grains, i.e. as *starch* and as refined sugar, *sucrose* in food sweetenings and preservatives.

Digested in duodenum and ileum and absorbed as *glucose.*

Lipids

Highly reduced and therefore can be oxidized to release *energy.* Also important in *cell membranes* and as a component of *steroid hormones.*

Common sources: meat and animal foods are rich in *saturated fats* and *cholesterol,* plant sources such as sunflower and soya are rich in *unsaturated fats.*

Digested in duodenum and ileum and absorbed as *fatty acids and glycerol.*

Vitamins have no common structure or function but are essential in small amounts to use other dietary components efficiently. *Fat-soluble vitamins* (e.g. A, D and E) are ingested with fatty foods and *water-soluble vitamins* (B group, C) are common in fruits and vegetables.

Fibre (originally known as *roughage*) is mainly cellulose from plant cell walls and is common in fresh vegetables and cereals. It *may* provide some energy but mainly serves to aid faeces formation, prevent constipation and ensure the continued health of the muscles of large intestine.

An adequate diet provides sufficient *energy* for the performance of metabolic work, although the 'energy food' is in unspecified form.

A balanced diet provides all dietary requirements *in the correct proportions.* Ideally this would be *$^1/_7$ fat, $^1/_7$ protein* and *$^5/_7$ carbohydrate.*

In conditions of *undernutrition* the first concern is usually provision of an *adequate diet,* but to avoid symptoms of *malnutrition* a *balanced diet* must be provided.

Proteins are *building blocks* for growth and repair of many body tissues (e.g. myosin in muscle, collagen in connective tissues), as *enzymes,* as *transport systems* (e.g. haemoglobin), as *hormones* (e.g. insulin) and as *antibodies.*

Common source: meat, fish, eggs and legumes/pulses. Must contain eight *essential amino acids* since humans are not able to synthesize them. Animal sources generally contain more of the essential amino acids.

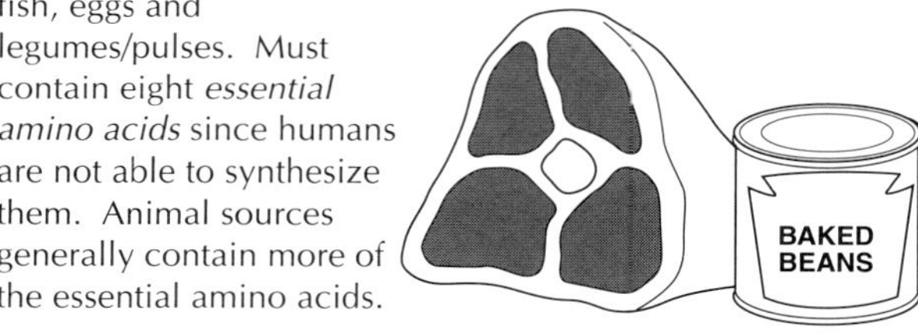

Digested in stomach, duodenum and ileum and absorbed as *amino acids.*

Water is required as a solvent, a transport medium, a substrate in hydrolytic reactions and for lubrication. A human requires 2-3 dm^3 of water daily, most commonly from drinks and liquid foods.

Minerals have a range of *specific* roles (direct structural components, e.g. Ca^{2+} in bones; constituents of macromolecules, e.g. PO_4^{3-} in DNA; part of pumping systems, e.g. Na^+ in glucose uptake; enzyme cofactors, e.g. Fe^{3+} in catalase; electron transfer, e.g. Cu^{2+} in cytochromes) and *collectively* help to maintain solute concentrations essential for control of water movement. They are usually ingested with other foods – dairy products and meats are particularly important sources.

Human digestive system: I

Palate: separates breathing and feeding pathways, allowing both processes to go on simultaneously so different types of teeth evolved.

Teeth: cut, tear and grind food so that solid foods are reduced to smaller particles for swallowing, and the food has a larger surface area for enzyme action.

Salivary glands: produce *saliva,* which is 99% water plus mucin, chloride ions (activate amylase), hydrogen carbonate and phosphate (maintain pH about 6.5), lysozyme and salivary amylase.
- **Paratoid**
- **Sublingual**
- **Submandibular**

Tongue: manoeuvres food for chewing and rolls food into a bolus for swallowing. Mixes food with saliva.

Diaphragm: a muscular 'sheet' separating the thorax and abdomen.

Liver: an accessory organ which produces bile and stores it in the gall bladder.

Bile duct: carries bile from gall bladder to duodenum.

Pancreas: an accessory organ producing a wide range of digestive secretions, as well as hormones.

Duodenum: the first 30 cm of the small intestine. Receives pancreatic secretions and bile and produces an alkaline mucus for protection, lubrication and chyme neutralization.

Ileum: up to 6 m in length – main site for absorption of soluble products of digestion.

Appendix: no function in humans. It is a vestige of the caecum in other mammals (herbivores).

Anal sphincter: regulates release of faeces (defecation).

Uvula: extension of soft palate which separates nasal chamber from pharynx.

Epiglottis: muscular flap which reflexly closes the trachea during swallowing to prevent food entry to respiratory tree.

Oesophagus: muscular tube which is dorsal to the trachea and connects the buccal cavity to the stomach. Muscular to generate peristaltic waves, which drive bolus of food downwards, and glandular to lubricate bolus with mucus. Semi-solid food passes to stomach in 4–8 seconds, very soft foods and liquids take only 1 second.

Cardiac sphincter: allows entry of food to stomach. Helps to retain food in stomach.

Stomach: a muscular bag which is distensible to permit storage of large quantities of food. Mucosal lining is glandular, with numerous gastric pits which secrete digestive juices. Three muscle layers including an oblique layer churn the stomach contents to ensure thorough mixing and eventual transfer of chyme to the duodenum.

Pyloric sphincter: opens to permit passage of chyme into duodenum and closes to prevent backflow of food from duodenum to stomach.

Colon (large intestine): absorbs water from faeces. Some B vitamins and vitamin K are synthesized by colonic bacteria. Mucus glands lubricate faeces.

Rectum: stores faeces before expulsion.

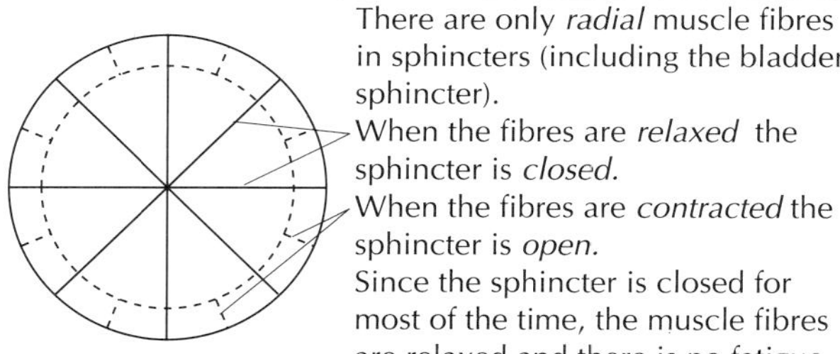

There are only *radial* muscle fibres in sphincters (including the bladder sphincter).
When the fibres are *relaxed* the sphincter is *closed.*
When the fibres are *contracted* the sphincter is *open.*
Since the sphincter is closed for most of the time, the muscle fibres are relaxed and there is no fatigue.

Human digestive system: II

Digestion of protein, fat and carbohydrate

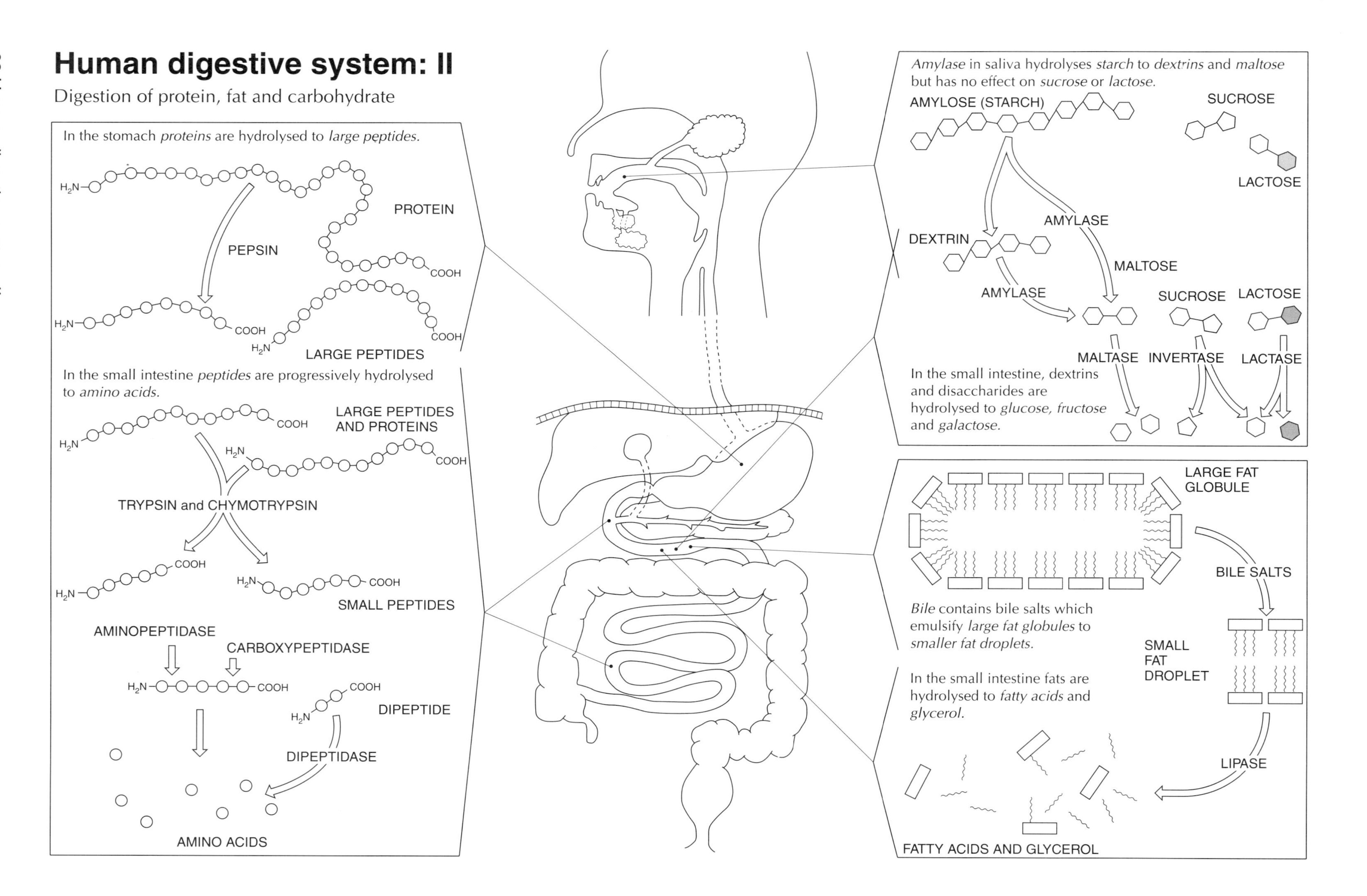

Absorption of the products of digestion

Absorption of the products of digestion is aided by a large surface area, specific uptake systems and well-developed transport network.

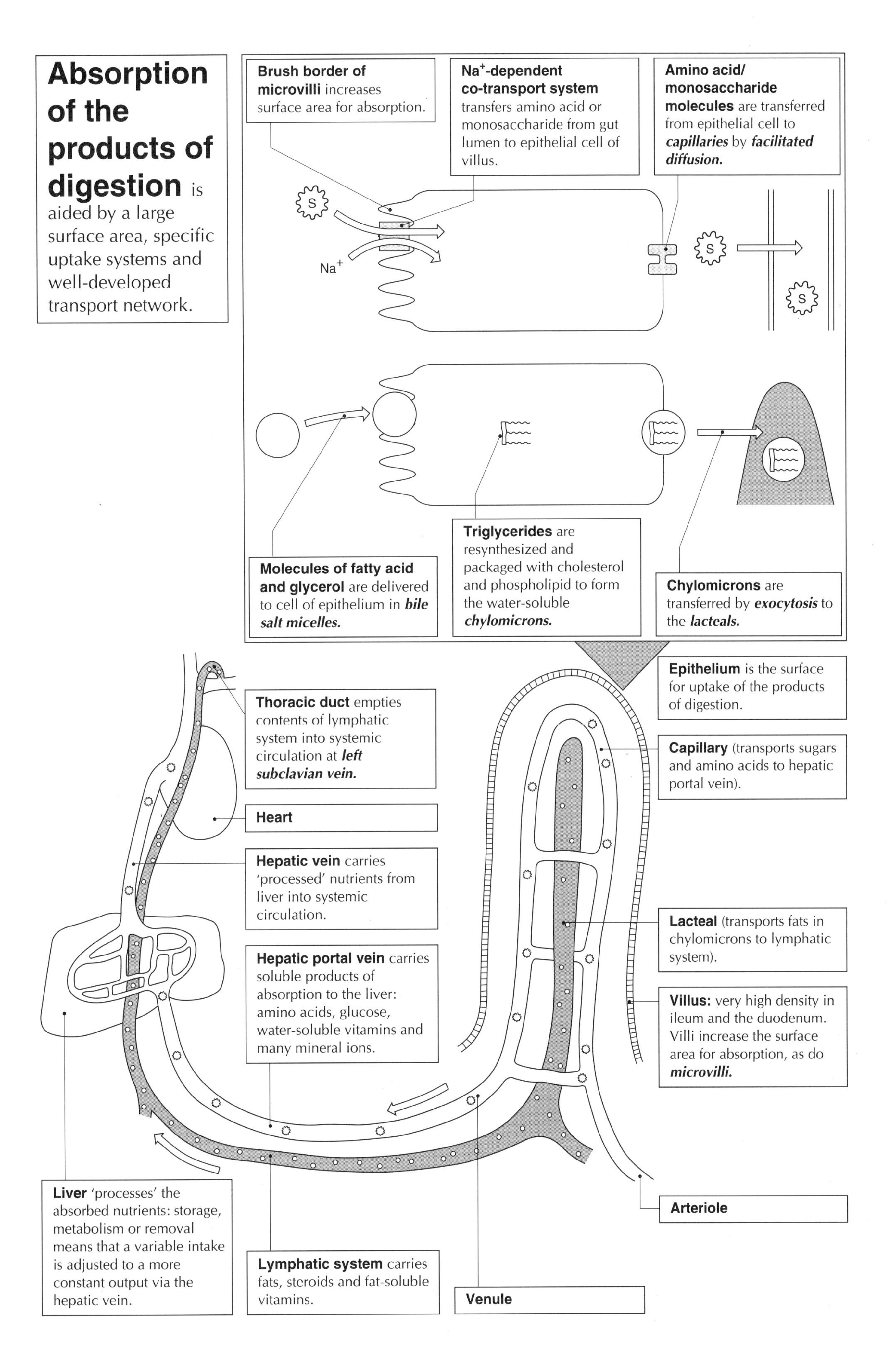

Principles of respiration:

a number of processes are involved in the provision/consumption of ***oxygen*** and the excretion/production of ***carbon dioxide.***

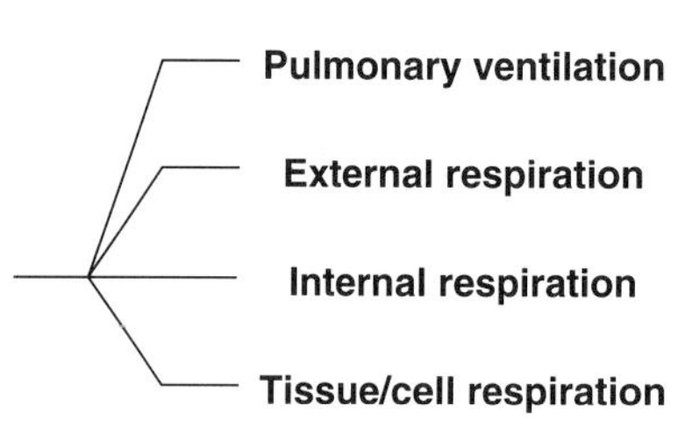

- **Pulmonary ventilation** moves gases between atmosphere and respiratory surface.
- **External respiration** occurs when gases diffuse across the respiratory surface.
- **Internal respiration** occurs when gases diffuse between circulating blood and respiring cells.
- **Tissue/cell respiration** occurs when oxygen is consumed and carbon dioxide is produced during the oxidation of foods to release energy.

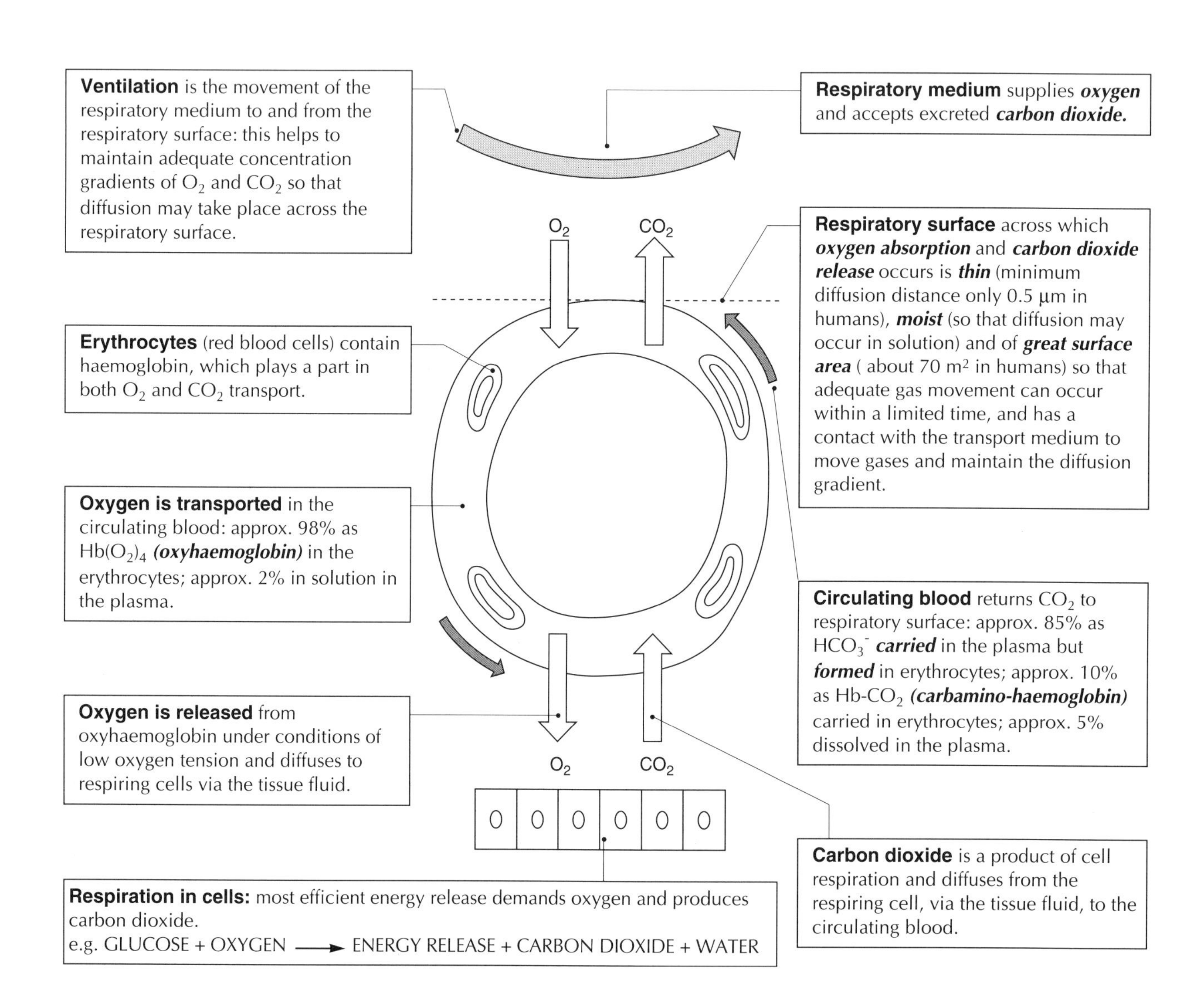

FEATURE	INSECT	FISH	MAMMAL
RESPIRATORY MEDIUM	AIR	WATER	AIR
SURFACE	TRACHEOLES contact cells directly	LAMELLAE of gills	ALVEOLI of lungs
TRANSPORT SYSTEM	NONE	Blood in SINGLE CIRCULATION	Blood in DOUBLE CIRCULATION
VENTILATION	Little – some ABDOMINAL MOVEMENT	Muscular movements drive ONE-WAY flow of water	Negative pressure system initiates TIDAL FLOW of air

Lung structure and function may be affected by a variety of disease conditions.

Larynx containing vocal cords which snap shut during hiccoughs.

Laryngitis is an inflammation of the larynx caused by a viral infection, often followed by a secondary bacterial infection.

Trachea with C-shaped rings of cartilage to support trachea in open position when thoracic pressure falls.

Lining of ciliated epithelium to trap and remove pathogens and particles of dust and smoke.

External intercostal muscles which contract to lift rib cage upwards and outwards.

Pleural membranes are moist to reduce friction as lungs move in chest.

Heart is close to lungs to drive pulmonary circulation.

Cut end of rib: ribs protect lungs and heart.

Pleural cavity at negative pressure so that passive lungs follow movement of rib cage.

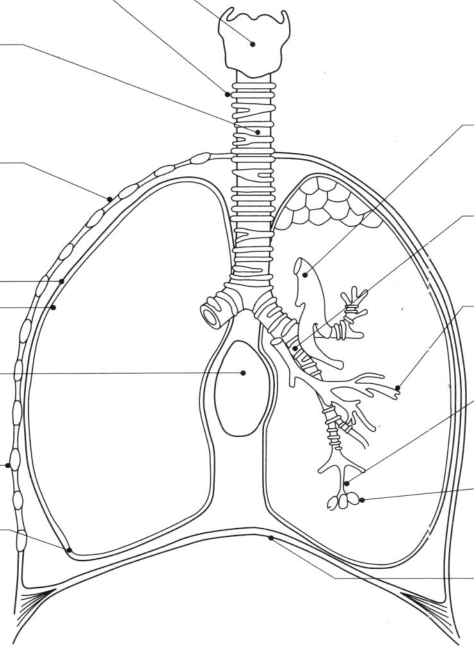

Chronic bronchitis is a progressive inflammatory disease caused by exposure to irritants, including tobacco smoke, sulphur dioxide and urban fog. The mucous membrane is damaged causing swelling and fluid secretion, and reduced ciliary activity allows excess mucus to collect. There may be difficulty in breathing, and bacteria may infect the stagnant mucus, leading to pus formation. Continued shortage of oxygen leads to pulmonary hypertension and death.

Pleurisy is infection of the pleural membranes, causing painful breathing and impairment of the negative pressure breathing system.

Epiglottis covers trachea when swallowing to prevent entry of food to lungs.

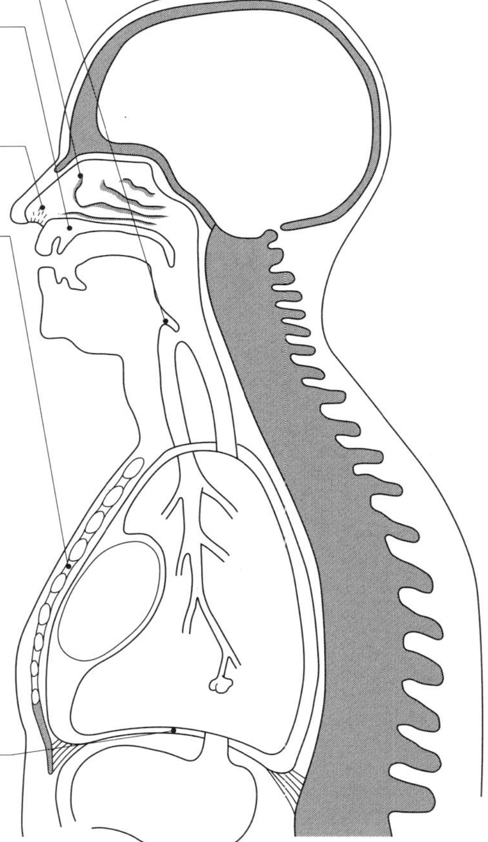

Turbinate bones direct flow of air so that inspired air is warmed and moistened.

Palate separates nasal and buccal cavities to permit breathing and feeding at the same time.

Nasal hairs trap dust particles and some air-borne pathogens.

Sternum for ventral attachment of ribs.

Pulmonary artery, which delivers deoxygenated blood to the lungs.

Bronchus with cartilaginous rings to prevent collapse during inspiration.

Pulmonary vein, which returns oxygenated blood to the heart.

Terminal bronchiole with no cartilage support.

Alveolus, which is the actual site of gas exchange between air and blood.

Domed muscular diaphragm, whose contraction may increase the volume of the thorax.

Emphysema is linked to tobacco smoking, which stimulates the release of proteolytic enzymes from mast cells in the lungs. These enzymes break down alveolar walls producing single large chambers. Thus the effective surface area of the lungs is decreased, causing reduced oxygenation of the blood.

Lung cancer arises from a tumour which develops in the bronchus and then invades adjacent tissues. It causes loss of function and pain, and tumour cells may spread via the bloodstream to other parts of the body.

Fine structure of the lung: exchange of gases in the alveolus requires

1. A tube for the movement of gases to and from the atmosphere (e.g. ***bronchiole).***
2. A surface across which gases may be transported between air and blood (i.e. ***alveolar membrane).***
3. A vessel which can take away oxygenated blood or deliver carboxylated blood (i.e. a ***branch of the pulmonary circulation).***

Inspired air

Expired air

Terminal bronchiole has no rings of cartilage and collapses when external pressure is high – dangerous when diving as trapped air in alveolus may give up nitrogen to blood, where it forms damaging bubbles.

Alveolar duct (atrium)

Alveolus (air sac)

Elastic fibres in alveolus permit optimum extension during inspiration – properties are adversely affected by tobacco smoke → ***emphysema.***

Branch of pulmonary artery delivers deoxygenated blood to the alveolar capillaries.

Surfactant is a phospholipid produced by ***septal cells*** in the alveolar wall. It reduces surface tension of the alveolar walls and prevents them sticking together following expiration - its absence in the newborn may lead to ***respiratory distress syndrome,*** and even to death, since the effort needed to breathe is 7–10 × normal.

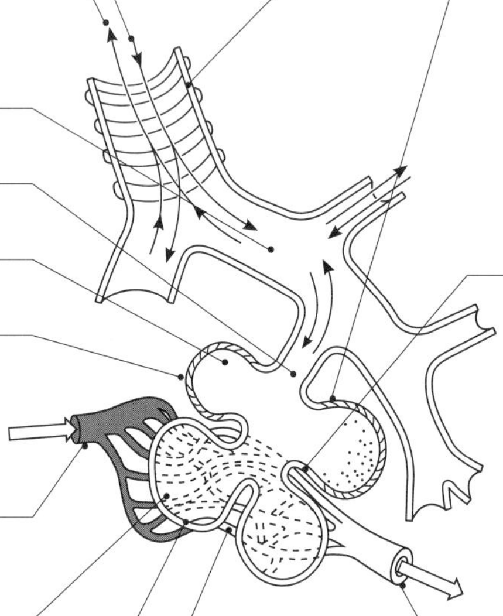

Bronchiole has supporting rings of cartilage to prevent collapse during low pressure phase of breathing cycle.

P_{450} is a cytochrome which speeds oxygen transfer across the alveolar membrane by facilitating diffusion, and is also involved in the detoxification of some harmful compounds by oxidation. Toxins in tobacco smoke may drive P_{450} to ***consume*** O_2 rather than ***transport*** it, leading to ***anoxia*** (oxygen deficiency).

Alveolar-capillary (respiratory) membrane consists of
1. ***Alveolar wall:*** squamous epithelium and alveolar macrophages.
2. ***Epithelial*** and ***capillary basement membranes.***
3. ***Endothelial cells of the capillary wall.***

Despite the number of layers this membrane averages only 0.5 μm in thickness.

Tributary of pulmonary vein returns oxygenated blood to the four pulmonary veins and thence to the left atrium of the heart.

Alveolar capillaries adjacent to the alveolus are the site of oxygen and carbon dioxide transfer between the air in the alveolus (air sac) and the circulating blood.

Stretch receptors provide sensory input, which initiates the ***Hering-Breuer reflex*** control of the breathing cycle.

Changes in composition of inspired and expired air

	INSPIRED	ALVEOLAR	EXPIRED	
O_2	20.95	13.80	16.40	Oxygen diffuses from alveoli into blood: expired air has an increased proportion of oxygen due to additional oxygen added from the anatomical dead space.
CO_2	0.04	5.50	4.00	Carbon dioxide concentration in alveoli is high because CO_2 diffuses from blood: the apparent fall in CO_2 concentration in expired air is due to dilution in the anatomical dead space.
N_2	79.01	80.70	79.60	The apparent increase in the concentration of nitrogen, a metabolically inert gas, is due to a ***relative*** decrease in the proportion of oxygen rather than an ***absolute*** increase in nitrogen.
$H_2O(g)$	VARI–ABLE	SATURATED		The moisture lining the alveoli evaporates into the alveolar air and is then expired unless the animal has anatomical adaptations to prevent this (e.g. the extensive nasal hairs in desert rats).
Temp.	ATMOS–PHERIC	BODY		Heat lost from the blood in the pulmonary circulation raises the temperature of the alveolar air.

Pulmonary ventilation is a result of changes in pressure within the thorax.

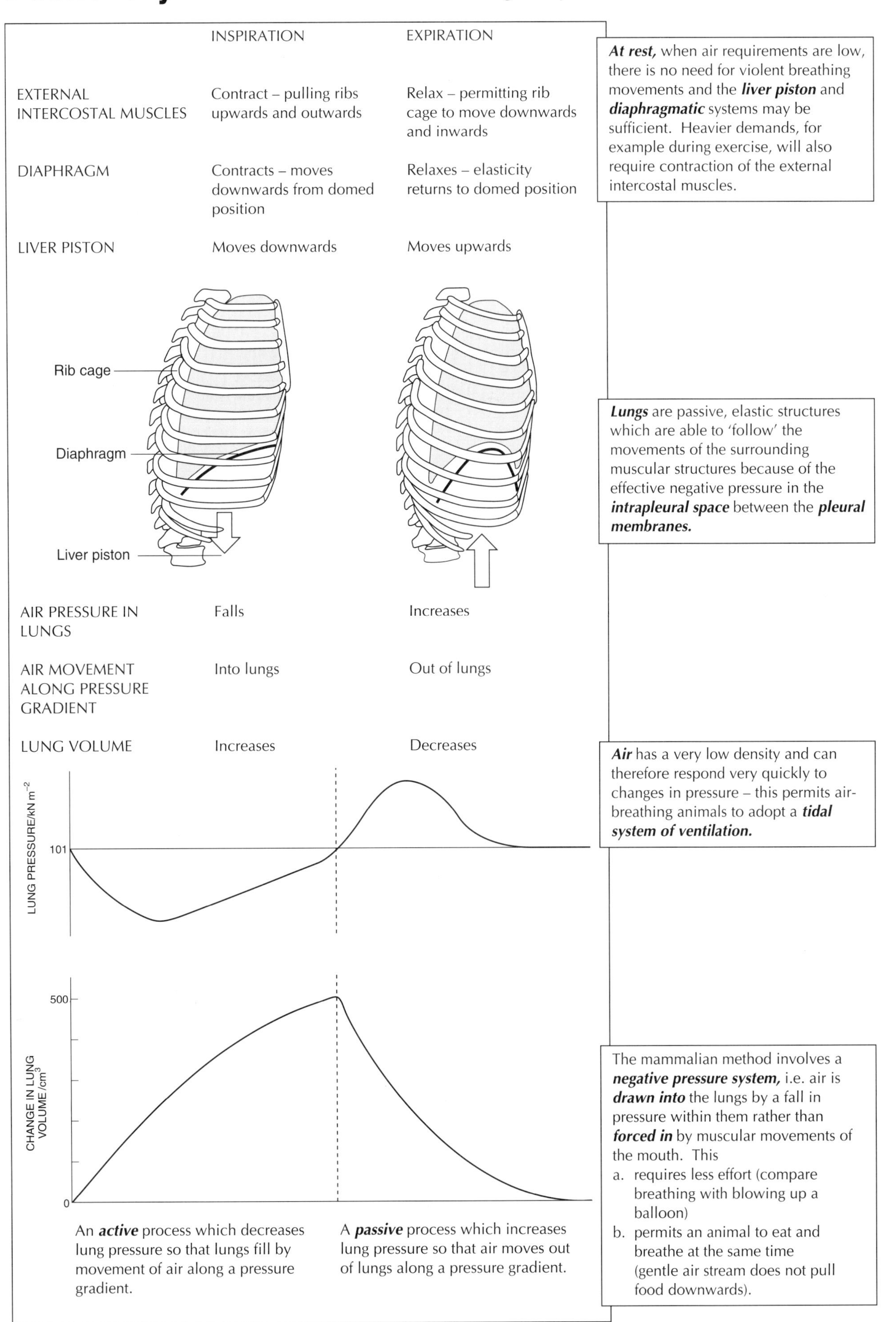

	INSPIRATION	EXPIRATION
EXTERNAL INTERCOSTAL MUSCLES	Contract – pulling ribs upwards and outwards	Relax – permitting rib cage to move downwards and inwards
DIAPHRAGM	Contracts – moves downwards from domed position	Relaxes – elasticity returns to domed position
LIVER PISTON	Moves downwards	Moves upwards
AIR PRESSURE IN LUNGS	Falls	Increases
AIR MOVEMENT ALONG PRESSURE GRADIENT	Into lungs	Out of lungs
LUNG VOLUME	Increases	Decreases

An ***active*** process which decreases lung pressure so that lungs fill by movement of air along a pressure gradient.

A ***passive*** process which increases lung pressure so that air moves out of lungs along a pressure gradient.

At rest, when air requirements are low, there is no need for violent breathing movements and the ***liver piston*** and ***diaphragmatic*** systems may be sufficient. Heavier demands, for example during exercise, will also require contraction of the external intercostal muscles.

Lungs are passive, elastic structures which are able to 'follow' the movements of the surrounding muscular structures because of the effective negative pressure in the ***intrapleural space*** between the ***pleural membranes.***

Air has a very low density and can therefore respond very quickly to changes in pressure – this permits air-breathing animals to adopt a ***tidal system of ventilation.***

The mammalian method involves a ***negative pressure system,*** i.e. air is ***drawn into*** the lungs by a fall in pressure within them rather than ***forced in*** by muscular movements of the mouth. This

a. requires less effort (compare breathing with blowing up a balloon)
b. permits an animal to eat and breathe at the same time (gentle air stream does not pull food downwards).

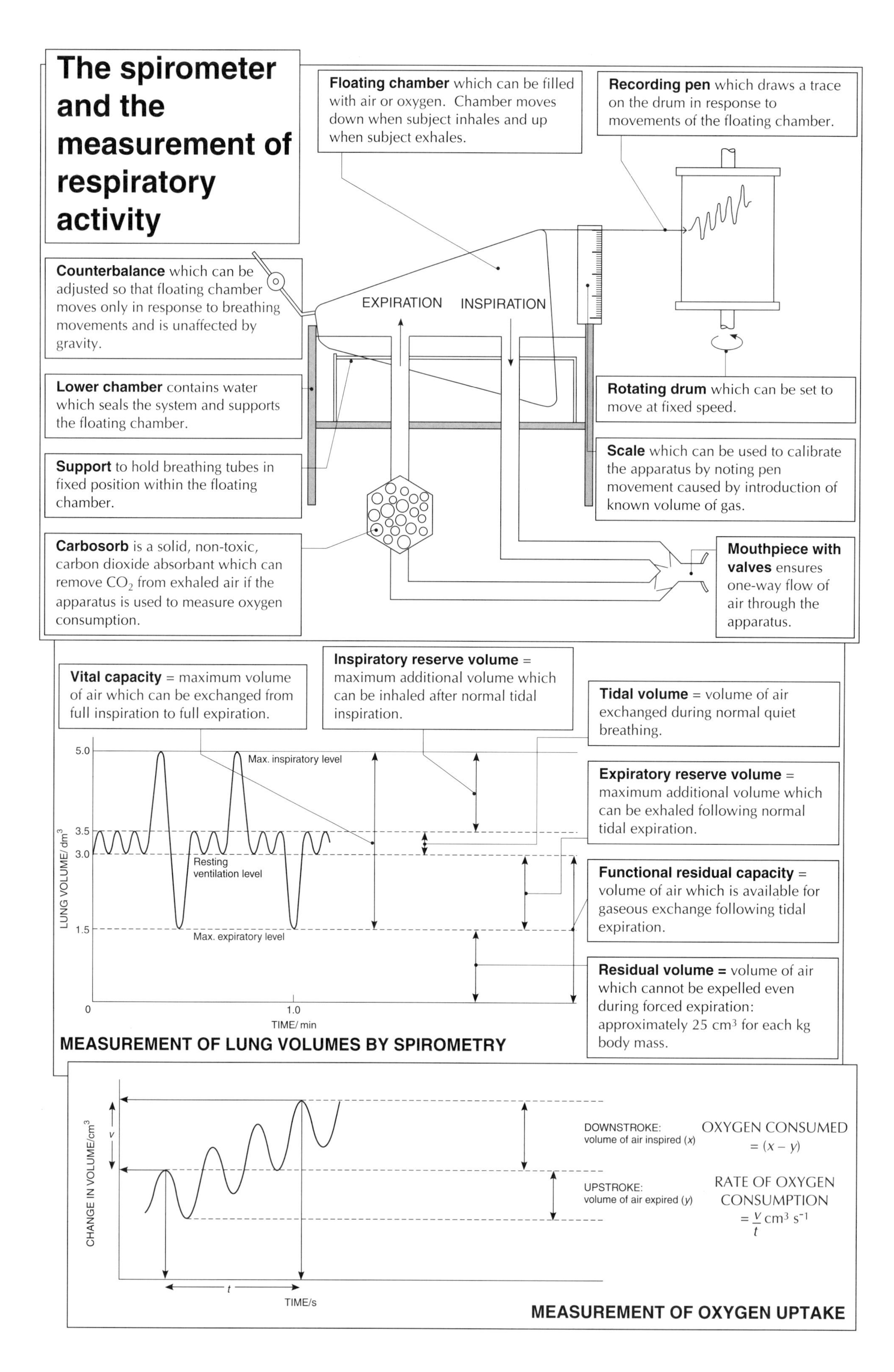
The spirometer and the measurement of respiratory activity
Floating chamber which can be filled with air or oxygen. Chamber moves down when subject inhales and up when subject exhales.
Recording pen which draws a trace on the drum in response to movements of the floating chamber.
Counterbalance which can be adjusted so that floating chamber moves only in response to breathing movements and is unaffected by gravity.
EXPIRATION
INSPIRATION
Lower chamber contains water which seals the system and supports the floating chamber.
Rotating drum which can be set to move at fixed speed.
Support to hold breathing tubes in fixed position within the floating chamber.
Scale which can be used to calibrate the apparatus by noting pen movement caused by introduction of known volume of gas.
Carbosorb is a solid, non-toxic, carbon dioxide absorbant which can remove CO_2 from exhaled air if the apparatus is used to measure oxygen consumption.
Mouthpiece with valves ensures one-way flow of air through the apparatus.
Vital capacity = maximum volume of air which can be exchanged from full inspiration to full expiration.
Inspiratory reserve volume = maximum additional volume which can be inhaled after normal tidal inspiration.
Tidal volume = volume of air exchanged during normal quiet breathing.
Expiratory reserve volume = maximum additional volume which can be exhaled following normal tidal expiration.
Functional residual capacity = volume of air which is available for gaseous exchange following tidal expiration.
Residual volume = volume of air which cannot be expelled even during forced expiration: approximately 25 cm³ for each kg body mass.
5.0
3.5
3.0
1.5
0
1.0
Max. inspiratory level
Resting ventilation level
Max. expiratory level
LUNG VOLUME/ dm³
TIME/ min
MEASUREMENT OF LUNG VOLUMES BY SPIROMETRY
CHANGE IN VOLUME/cm³
v
t
TIME/s
DOWNSTROKE: volume of air inspired (x)
UPSTROKE: volume of air expired (y)
OXYGEN CONSUMED = (x – y)
RATE OF OXYGEN CONSUMPTION = v/t cm³ s⁻¹
MEASUREMENT OF OXYGEN UPTAKE

Blood cells differ in structure and function.

If blood is spun for a few minutes in a high speed centrifuge it separates into two layers.

Serum is the name given to plasma from which the soluble protein fibrinogen (a protein involved in blood clotting) has been removed.

Blood cells originate from *stem cells* in the bone marrow by the process of *haemopoiesis*.

Erythrocytes (red blood cells) are the most numerous of blood cells – about 5 000 000 per mm^3 of adult blood. They function in the transport of O_2 and CO_2, and they contribute to the buffering capacity of the blood. The red colour is due to the presence of the pigment haemoglobin. There are several advantages in packing the haemoglobin into cells rather than leaving it free in the cytoplasm – it keeps the blood viscosity low, it allows the best arrangement of enzymes and solutes for functioning of haemoglobin and it prevents a dramatic reduction in blood water potential. The typical lifespan of a red cell is 90–120 days, before they are destroyed in the spleen.

Platelets (thrombocytes) are fragments of cells which are involved in blood clotting (they disintegrate to release thromboplasts).

Neutrophils are the most abundant of the leucocytes (white blood cells). They are very short-lived (12–72 h), contain non-staining granules, and are responsible for the phagocytosis of micro-organisms. They migrate from the blood to the tissues, and are so active in phagocytosis that they are replaced at the rate of about 100 000 000 000 per day.

Monocytes are the largest of the leucocytes. They are agranulocytes (have non-granular cytoplasm) and have a large, bean-shaped nucleus. They spend a short time (2–3 days) in the circulatory system before moving into the tissues where they mature into phagocytic macrophages.

Eosinophils have a double-lobed nucleus and granules which stain red with the acid dye eosin. They help control the allergic response – for example, they secrete enzymes which inactivate histamine. Their numbers increase during allergic reactions and in response to some parasitic infections (e.g. tapeworm and hookworm).

Lymphocytes make up about 30% of the circulating leucocytes. Although they are produced in the bone marrow they continue to develop and mature in the lymph nodes, the thymus gland and the spleen. They are responsible for the specific immune response – the B-lymphocytes produce antibodies and the T-lymphocytes have a number of roles, including co-ordination of the immune response and direct cell destruction. They are best identified by the prominent, deeply staining nucleus and the thin 'halo' of clear cytoplasm.

Basophils have an S-shaped nucleus and granules which stain blue. They secrete large amounts of histamine (which increases inflammation) and heparin (which helps to keep a balance between blood clotting and not clotting).

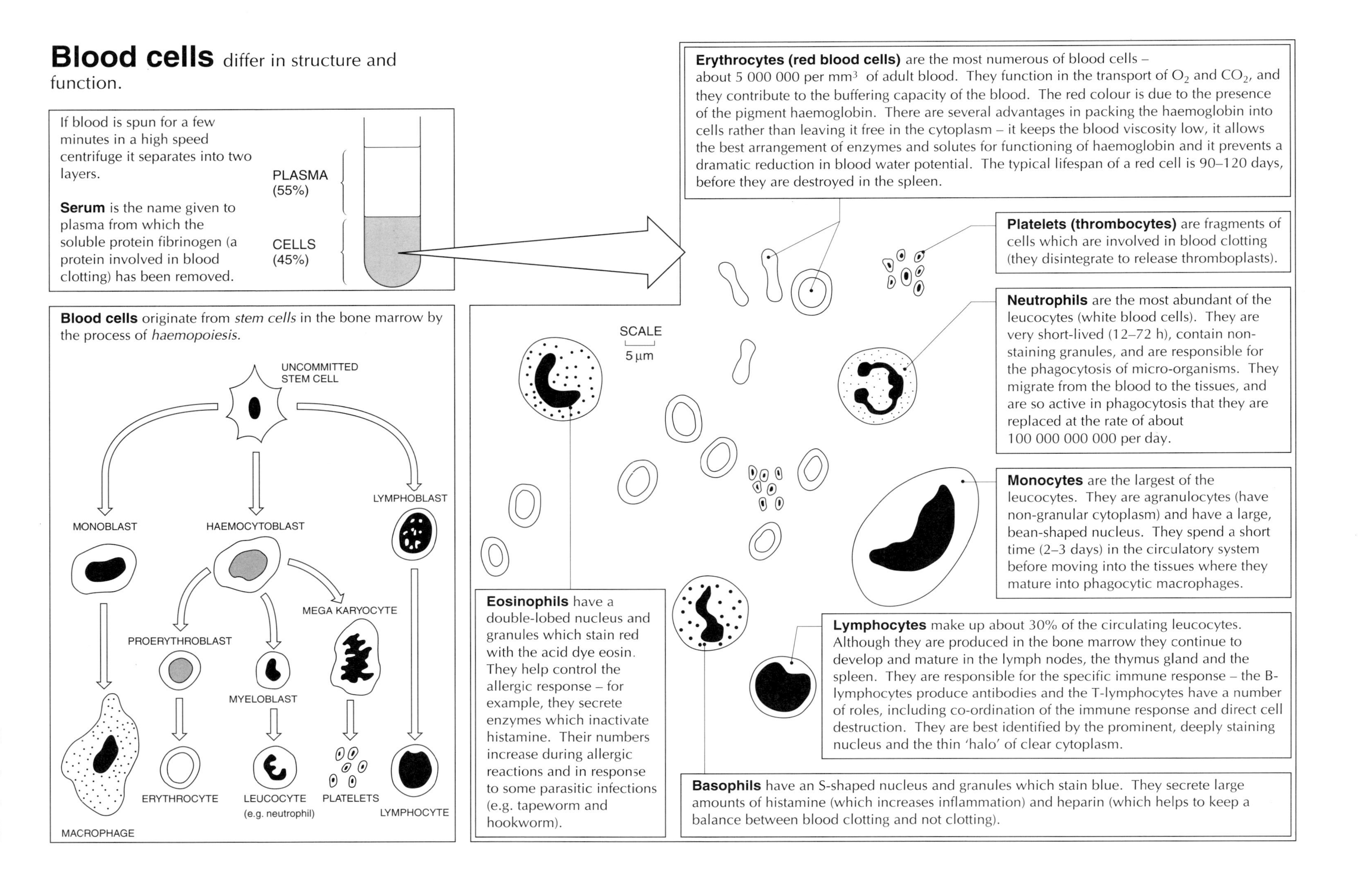

Tissue fluid (interstitial or intercellular fluid)

Tissue fluid (interstitial or intercellular fluid) is the immediate environment of the cells, and represents the 'internal environment' described by Claude Bernard in his definition of homeostasis.

Plasma proteins do not move from plasma to tissue fluid (cannot cross capillary endothelium) - largely responsible for solute potential of plasma.

Movement from tissue fluid to plasma
- ***Water***
- ***Carbon dioxide***
- ***Nitrogenous waste***
- ***Hormones and other secretions***

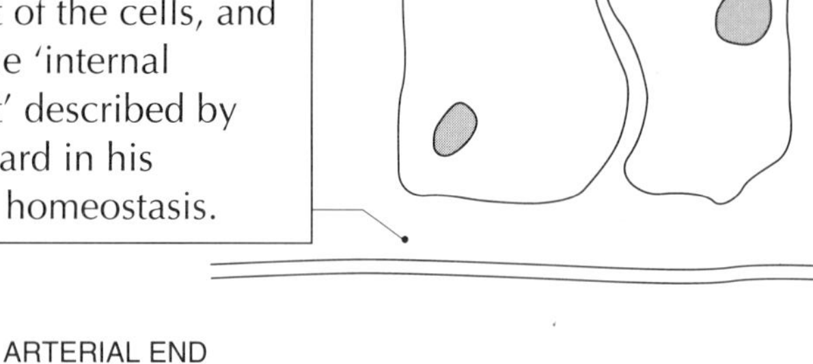

Movement from plasma to tissue fluid
- ***Water***
- ***Oxygen***
- ***Soluble products of digestion***
- ***Hormones***

Living cells place demands on the tissue fluid.

FORCES WHICH REGULATE THE FORMATION AND RECLAMATION OF TISSUE FLUID

Net force driving fluid movement at any point = pressure potential gradient – solute potential gradient

Pressure potential (hydrostatic potential) is the pressure exerted on a fluid by its surroundings, e.g. by ***pumping action of heart*** and ***elastic recoil of arteries.***

Solute potential (osmotic potential) is the force of attraction towards water molecules caused by dissolved solutes, particularly ***ions*** and ***plasma proteins***.

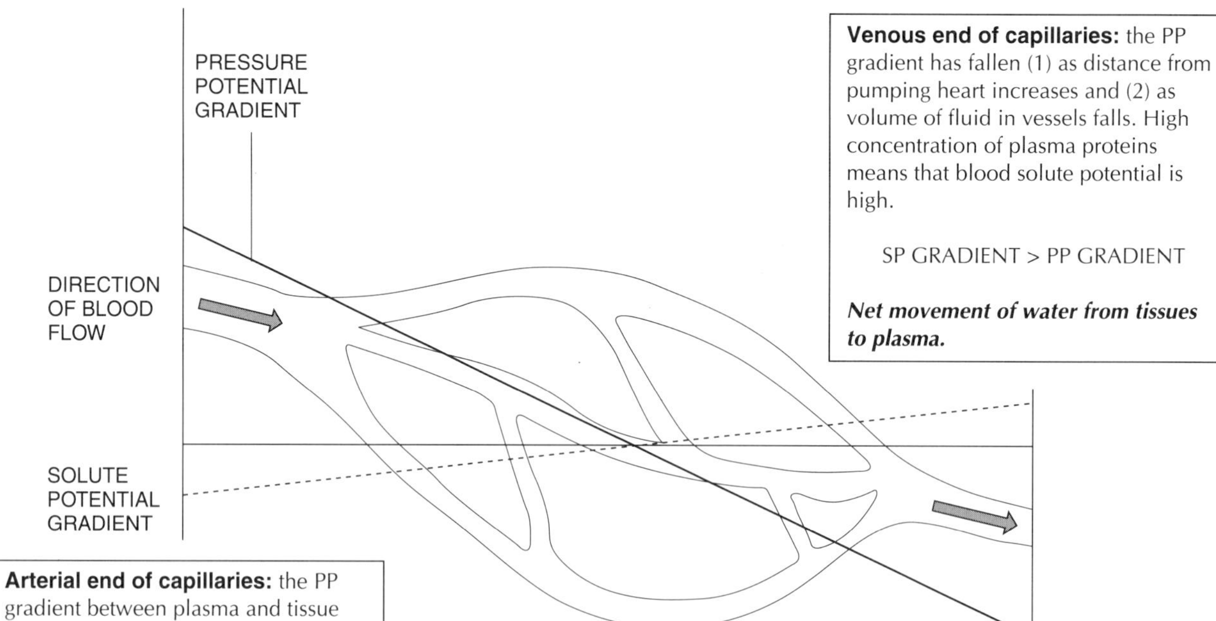

Venous end of capillaries: the PP gradient has fallen (1) as distance from pumping heart increases and (2) as volume of fluid in vessels falls. High concentration of plasma proteins means that blood solute potential is high.

SP GRADIENT > PP GRADIENT

Net movement of water from tissues to plasma.

Arterial end of capillaries: the PP gradient between plasma and tissue fluid is high due to pumping of heart and recoil of artery walls.

PP GRADIENT > SP GRADIENT

Net movement of water from plasma to tissue fluid.

In ***most capillaries*** there is a net flow of fluid from the blood to the tissue fluid. This depends on the pressure potential and solute potential gradients between blood plasma and tissue fluid – because the pressure potential falls as blood travels through the capillaries whereas blood solute potential remains fairly constant water tends to leave the capillaries at the high pressure end and enter at the low pressure end. Any net loss drains to the lymphatic system.

Functions of blood

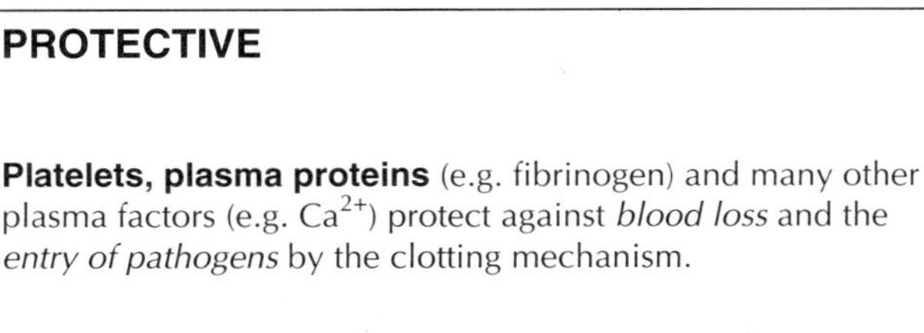

REGULATORY

Blood solutes affect the water potential of the blood, and thus the water potential gradient between the blood and the tissue fluid. The size of this water potential gradient, determined principally by plasma concentrations of Na^+ ions and plasma proteins, thus *regulates water movement* between blood and tissues.

The **water content** of the blood plays a part in *regulation of body temperature* since it may transfer heat between thermogenic (heat-generating) centres, such as the liver, skeletal muscle and brown fat, and heat sinks such as the skin, the brain and the kidney.

pH maintenance is an important function of *blood buffer systems* such as the hydrogencarbonate and phosphate equilibria, and is a secondary role of haemoglobin and some plasma proteins.

PROTECTIVE

Platelets, plasma proteins (e.g. fibrinogen) and many other plasma factors (e.g. Ca^{2+}) protect against *blood loss* and the *entry of pathogens* by the clotting mechanism.

Leucocytes protect against *toxins and potential pathogens* by both non-specific (e.g. phagocytosis) and specific (e.g. antibody production and secretion) immune responses.

TRANSPORT

Soluble products of digestion/absorption (such as glucose, amino acids, vitamins and minerals) from the gut to the liver and then to the general circulation. Fatty acids are transported from the gut to the lymph system and then to the general circulation.

Waste products of metabolism (such as urea, creatinine and lactate) from sites of production to sites of removal, such as the liver and kidney.

Hormones (such as insulin, a peptide, testosterone, a steroid, and adrenaline, a catecholamine) from their sites of production in the glands to the target organs where they exert their effects.

Respiratory gases (oxygen and carbon dioxide) from their sites of uptake or production to their sites of utilization or removal. Oxygen transport is more closely associated with red blood cells, and carbon dioxide transport with the plasma.

Plasma proteins secreted from the liver and present in the circulating blood include fibrinogen (a blood clotting agent), globulins (involved with specific transport functions, e.g. of thyroxine, iron and copper) and albumin (which binds plasma Ca^{2+} ions).

Haemoglobin and myoglobin

OXYGEN DISSOCIATION CURVES OF HAEMOGLOBIN show the relationship between haemoglobin saturation with oxygen (i.e. the percentage of haemoglobin in the form of oxyhaemoglobin) and the partial pressure of oxygen in the environment (the pO_2 or oxygen tension).

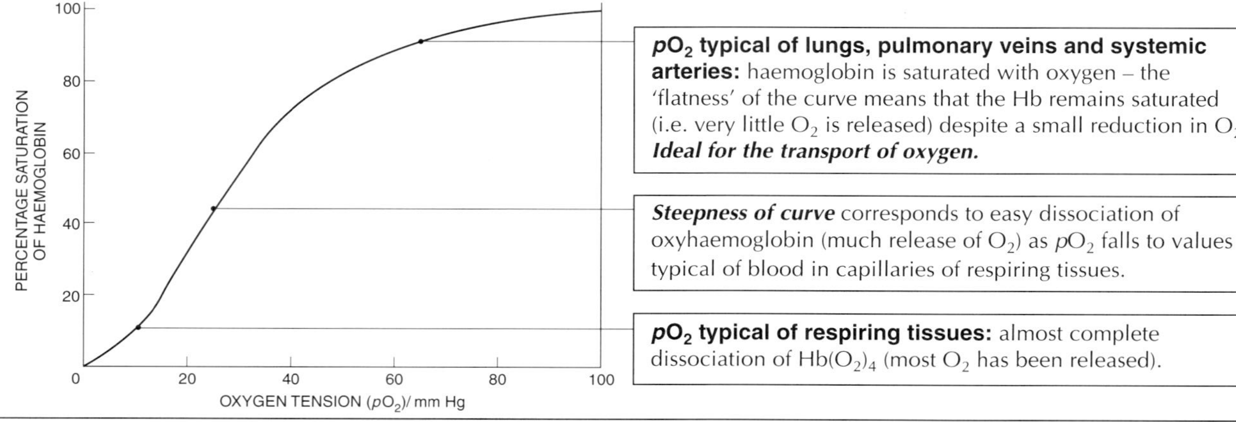

pO_2 typical of lungs, pulmonary veins and systemic arteries: haemoglobin is saturated with oxygen – the 'flatness' of the curve means that the Hb remains saturated (i.e. very little O_2 is released) despite a small reduction in O_2. ***Ideal for the transport of oxygen.***

Steepness of curve corresponds to easy dissociation of oxyhaemoglobin (much release of O_2) as pO_2 falls to values typical of blood in capillaries of respiring tissues.

pO_2 typical of respiring tissues: almost complete dissociation of $Hb(O_2)_4$ (most O_2 has been released).

The S-shape of the curve is most significant. Simply put, it means that oxygen associates with haemoglobin, and remains associated with it, at oxygen tensions typical of the alveolar capillaries, the pulmonary vein, the aorta and the arteries, but that it very rapidly dissociates from haemoglobin at oxygen tensions typical of those found in respiring tissues. Furthermore, this dissociation is almost complete at the low oxygen tensions found in the most active tissues. In other words, oxygen release from oxyhaemoglobin is tailored to the tissues' demand for this gas.

The reason for this S-shapedness is that haemoglobin and oxygen illustrate co-operative binding, that is, the binding of the first oxygen molecule to haemoglobin alters the shape of the haemoglobin molecule slightly so that the binding of a second molecule of oxygen is made easier, and so on until haemoglobin has its full complement of four molecules of oxygen. Conversely, when one molecule of oxygen dissociates from the oxyhaemoglobin, the haemoglobin shape is adjusted to make release of successive molecules of oxygen increasingly easy.

MYOGLOBIN, MUSCLE AND MARATHONS

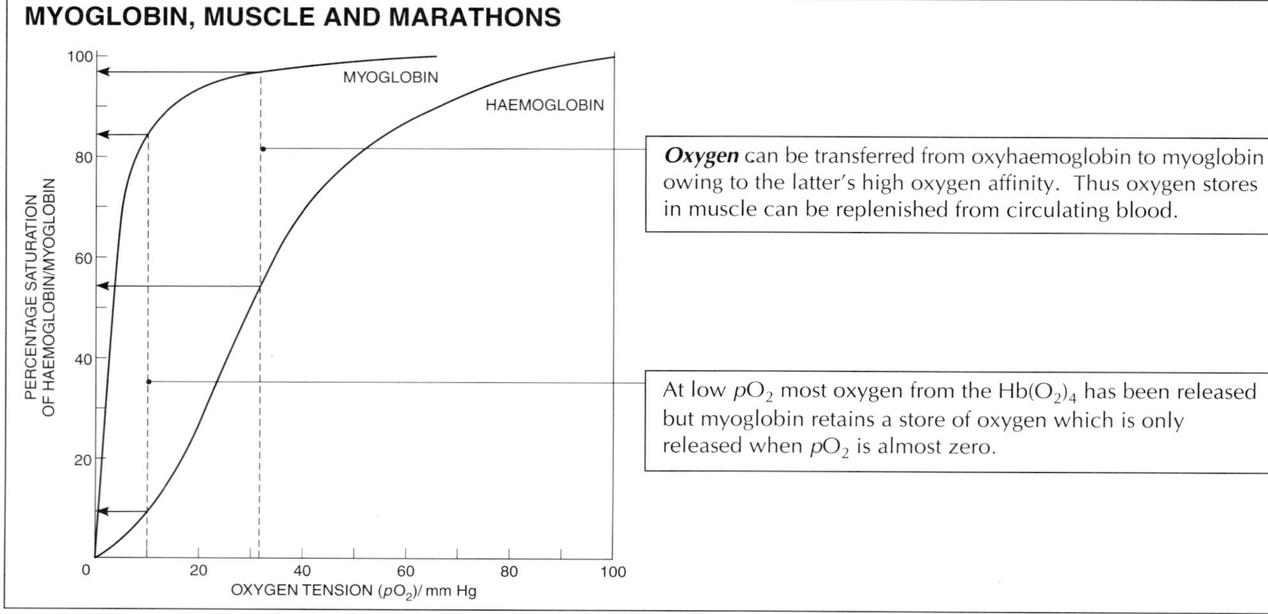

Oxygen can be transferred from oxyhaemoglobin to myoglobin owing to the latter's high oxygen affinity. Thus oxygen stores in muscle can be replenished from circulating blood.

At low pO_2 most oxygen from the $Hb(O_2)_4$ has been released but myoglobin retains a store of oxygen which is only released when pO_2 is almost zero.

The muscles of mammals contain a red pigment called ***myoglobin*** which is structurally similar to one of the four sub-units of haemoglobin. This pigment may also bind to oxygen, but since there is only one haem group there can be no co-operative binding and the myoglobin–oxygen dissociation curve is hyperbolic rather than sigmoidal.

The significance of this is that at any particular oxygen tension myoglobin has a higher affinity for oxygen than does haemoglobin. Thus when oxyhaemoglobin in blood passes through tissues such as muscle, oxygen is transferred to the myoglobin. Further analysis of the myoglobin–oxygen dissociation curve will show that myoglobin does not release oxygen until oxygen tension is very low indeed, and myoglobin therefore represents an excellent ***store*** of oxygen.

Muscles which have a high oxygen demand during exercise, or which may be exposed to low oxygen tension in the circulating blood, commonly have particularly large myoglobin stores and are called 'red' muscles. A high proportion of red muscle is of considerable advantage to marathon runners, who must continue to respire efficiently even when their blood is severely oxygen-depleted.

The transport of carbon dioxide from tissue to lung

The red cell and haemoglobin both play a significant part in this process as well as in the transport of oxygen.

At the respiring tissue: carbon dioxide produced in the mitochondria diffuses out of the cells, through the plasma, and into the erythrocytes, where it combines with water to produce carbonic acid, H_2CO_3, under the influence of the enzyme carbonic anhydrase. This can dissociate into H^+ and HCO_3^-. (1)

The reaction proceeds rapidly since the equilibrium is disturbed by the rapid removal of the hydrogen ions (H^+) by association with haemoglobin to form haemoglobinic acid (H.Hb). By accepting hydrogen ions in this way haemoglobin is acting as a buffer, permitting the transport of large quantities of carbon dioxide without any significant change in blood pH. (2)

As a result of these changes the hydrogencarbonate concentration in the erythrocyte rises and these ions begin to diffuse along a concentration gradient into the plasma. However, this movement of negative ions is not balanced by an equivalent outward flow of positive ions since the membrane of the erythrocyte is relatively impermeable to sodium and potassium ions, which are therefore retained within the cell. This could potentially be disastrous, since positively charged erythrocytes would repel one another, a situation which would not enhance their function as oxygen carriers in the confines of a closed circulatory system! The situation is avoided, and electrical neutrality maintained, by an inward diffusion of chloride ions from the plasma sufficient to balance the HCO_3^- moving out. This movement of chloride ions to maintain erythrocyte neutrality is called the ***chloride shift.*** (3)

At the lungs: the reverse takes place. In the presence of oxygen haemoglobinic acid dissolves so that oxyhaemoglobin ($Hb.O_2$) may be formed. This releases hydrogen ions which combine with hydrogencarbonate from the plasma to produce carbonic acid. (4)

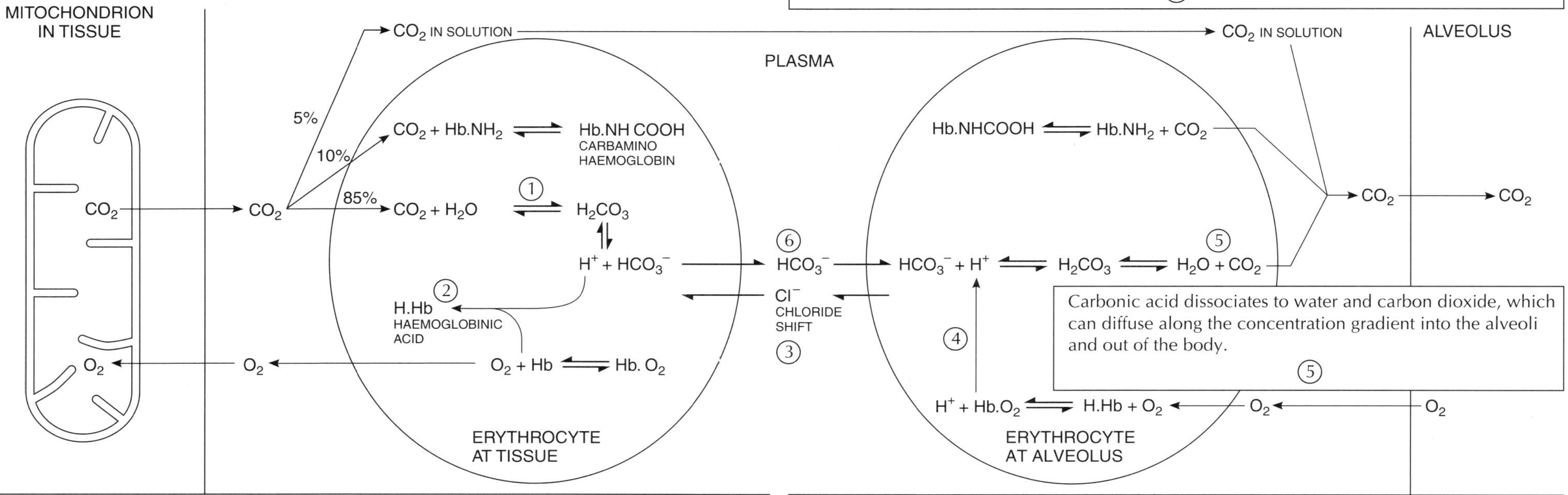

Carbonic acid dissociates to water and carbon dioxide, which can diffuse along the concentration gradient into the alveoli and out of the body. (5)

The hydrogencarbonate concentration in the red cells falls, more diffuses in from the plasma, and the process continues so that more carbon dioxide is released. Once more electrical neutrality of the erythrocytes is maintained by the chloride shift, but this time the chloride ions are moving in the opposite direction, from the cell to the plasma. (6)

MITOCHONDRION IN TISSUE

PLASMA

ALVEOLUS

ERYTHROCYTE AT TISSUE

CO_2 → CO_2: 5% → CO_2 IN SOLUTION; 10% → $CO_2 + Hb.NH_2 \rightleftharpoons Hb.NHCOOH$ (CARBAMINO HAEMOGLOBIN); 85% → $CO_2 + H_2O \rightleftharpoons H_2CO_3$ (1)

$H_2CO_3 \rightleftharpoons H^+ + HCO_3^-$

H^+ → H.Hb HAEMOGLOBINIC ACID (2)

$O_2 + Hb \rightleftharpoons Hb.O_2$; O_2 → O_2

(6) HCO_3^-; Cl^- CHLORIDE SHIFT (3)

ERYTHROCYTE AT ALVEOLUS

$Hb.NHCOOH \rightleftharpoons Hb.NH_2 + CO_2$

$HCO_3^- + H^+ \rightleftharpoons H_2CO_3 \rightleftharpoons H_2O + CO_2$ (5)

$H^+ + Hb.O_2 \rightleftharpoons H.Hb + O_2$ (4)

CO_2 IN SOLUTION → CO_2 → CO_2

O_2 → O_2

Mammalian double circulation comprises ***pulmonary*** (heart – lung – heart) and ***systemic*** (heart – rest of body – heart) ***circuits***. The complete separation of the two circuits permits rapid, high-pressure distribution of oxygenated blood essential in active, endothermic animals. There are many subdivisions of the systemic circuit, including ***coronary, cerebral, hepatic portal*** and, during fetal life only, ***fetal circuits***. The circuits are typically named for the organ or system which they service – thus each kidney has a ***renal*** artery and vein. Each organ has an artery bringing oxygenated blood and nutrients, and a vein removing deoxygenated blood and waste.

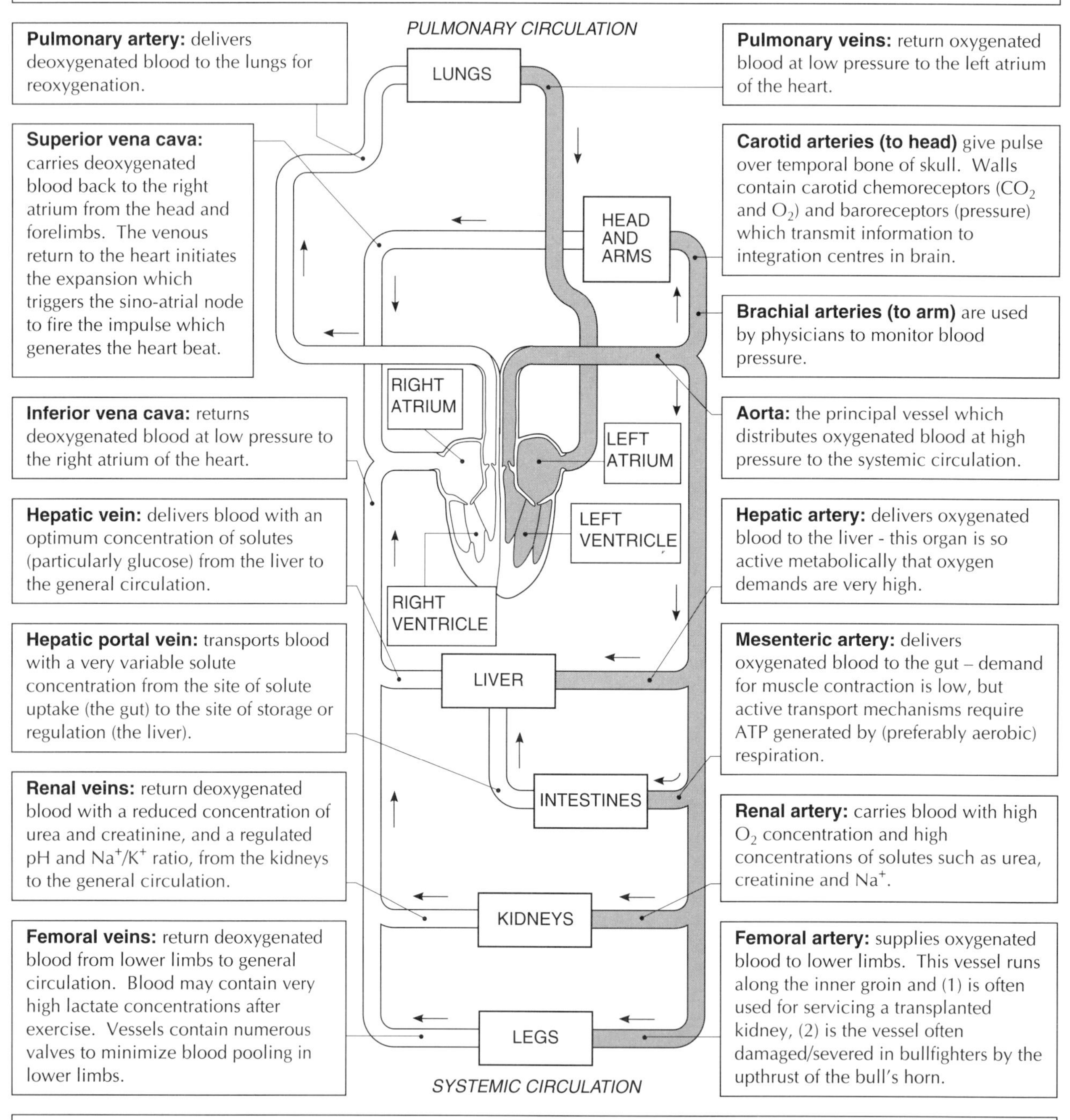

The **flow of blood** is maintained in three ways.

1. ***The pumping action of the heart:*** the ventricles generate pressures great enough to drive blood through the arteries into the capillaries.

2. ***Contraction of skeletal muscle:*** the contraction of muscles during normal movements compress and relax the thin-walled veins causing pressure changes within them. Pocket valves in the veins ensure that this pressure directs the blood to the heart, without backflow.

3. ***Inspiratory movements:*** reducing thoracic pressure caused by chest and diaphragm movements during inspiration helps to draw blood back towards the heart.

Mammalian heart: structure and function

Pulmonary (semilunar) valve: is composed of three cusps or watchpocket flaps which are forced together then the pressure in the pulmonary artery exceeds that in the right ventricle, thus preventing backflow of blood into the relaxing chambers of the heart.

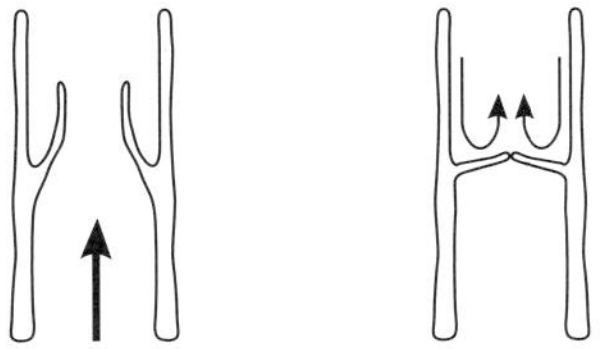

P ventricle>*P* artery

P ventricle<*P* artery

Tricuspid (right atrioventricular) valve: has three fibrous flaps with pointed ends which point into the ventricle. The flaps are pushed together when the ventricular pressure exceeds the atrial pressure so that blood is propelled past the inner edge of the valve through the pulmonary artery instead of through the valve and back into the atrium.

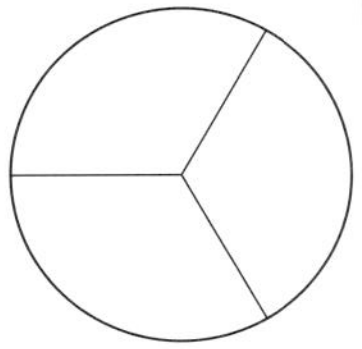

Superior (anterior) vena cava: carries deoxygenated blood back to the right atrium of the heart. As with other veins the wall is thin, with little elastic tissue or smooth muscle. In contrast to veins returning blood from below the heart there are no venous valves, since blood may return under the influence of gravity.

Control of heartbeat

1. The heartbeat is initiated in the ***sino-atrial node*** (particularly excitable myogenic tissue in the wall of the right atrium).
2. ***Intrinsic*** heart rate is about 78 beats per minute.
3. ***External (extrinsic)*** factors may modify basic heart rate:
   a. ***vagus nerve*** decreases heart rate;
   b. ***accelerator (sympathetic) nerve*** increases heart rate;
   c. ***adrenaline*** and ***thyroxine*** increase heart rate.

Resting heart rate of 70 beats per minute indicates that heart has ***vagal tone.***

Aorta: carries oxygenated blood from the left ventricle to the systemic circulation. It is a typical elastic (conducting) artery with a wall that is relatively thick in comparison to the lumen, and with more elastic fibres than smooth muscle. This allows the wall of the aorta to accommodate the surges of blood associated with the alternative contraction and relaxation of the heart - as the ventricles contract the artery expands and as the ventricle relaxes the elastic recoil of the artery forces the blood onwards.

Pulmonary arteries

Left atrium

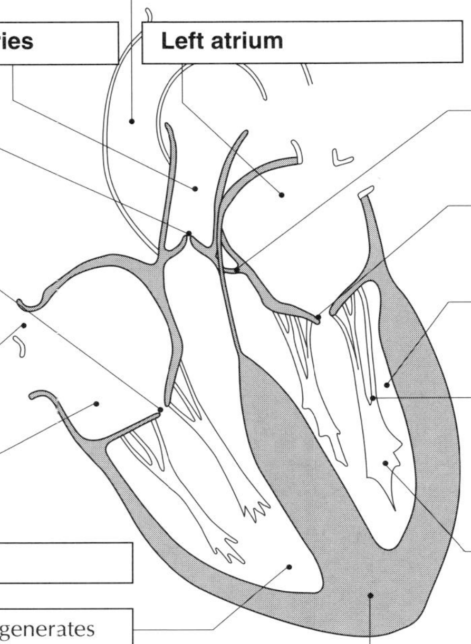

Right atrium

Right ventricle: generates pressure to pump deoxygenated blood to pulmonary circulation.

Myocardium is composed of cardiac muscle: intercalated discs separate muscle fibres, strengthen the muscle tissue and aid impulse conduction; cross-bridges promote rapid conduction throughout entire myocardium; numerous mitochondria permit rapid aerobic respiration. Cardiac muscle is myogenic (can generate its own excitatory impulse) and has a long refractory period (interval between two consecutive effective excitatory impulses), which eliminates danger of cardiac fatigue.

The pressure generated by the left ventricle is greater than that generated by the right ventricle as the systemic circuit is more extensive than the pulmonary circuit.

The pressure generated by the atria is less than that generated by the ventricles since the distance from atria to ventricles is less than that from ventricles to circulatory system.

Volume: the same volume of blood passes through each side of the heart. Both ventricles pump the same volume of blood.

Aortic (semilunar) valve: prevents backflow from aorta to left ventricle.

Bicuspid (mitral, left atrioventricular) valve: ensures blood flow from left ventricle into aortic arch.

Left ventricle: generates pressure to force blood into the systemic circulation.

Chordae tendinae: short, inextensible fibres – mainly composed of collagen – which connect to free edges of atrioventricular valves to prevent 'blow-back' of valves when ventricular pressure rises during contraction of myocardium.

Papillary muscles: contract as wave of excitation spreads through ventricular myocardium and tighten the chordae tendineae just before the ventricles contract.

CARDIAC MUSCLE FIBRE

CROSS BRIDGE WITH GAP JUNCTION

MITOCHONDRION

INTERCALATED DISC

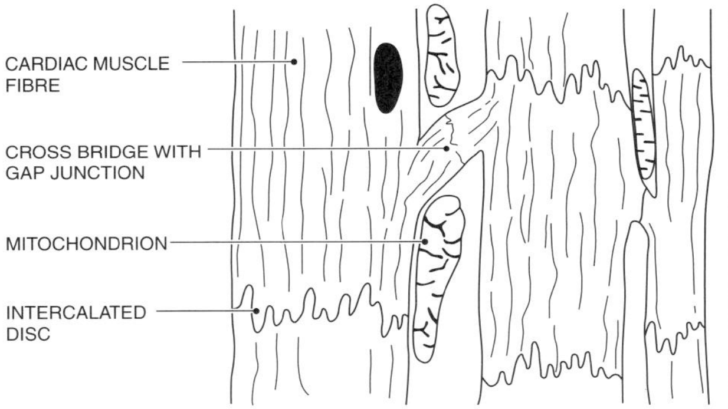

The lymphatic system

Tonsils are aggregations of large lymphatic nodules embedded in a mucous membrane. There is a single *pharyngeal* tonsil or *adenoid,* and two pairs (the *palatine* and the *lingual* tonsils), all arranged in a ring at the junction of the pharynx and the oral cavity. They contain phagocytes and lymphocytes which protect against the invasion of foreign substances from the mouth.

Right lymphatic duct receives lymph drainage from the right side of the head, the right upper trunk and the right arm, and empties the lymph into the systemic blood circulation at the junction of the right subclavian and right jugular veins.

Axillary nodes

The **thymus gland** is a paired organ most obvious in the pre-pubertal individual. It has a significant role in the immune response since it produces the T-lymphocytes which have a number of roles in the hierarchy of the immune system.

Peyer's patches are aggregated lymph nodes in the wall of the ileum, where they are ideally situated to offer protection against the invasion of potential pathogens or absorption of toxins from the gut contents.

The **lymphatic vessels** form a low-pressure return system for reclaimed tissue fluid. Since their contents are at low pressure the lymphatics are well-supplied with semi-lunar valves to ensure flow in one direction (towards the heart).

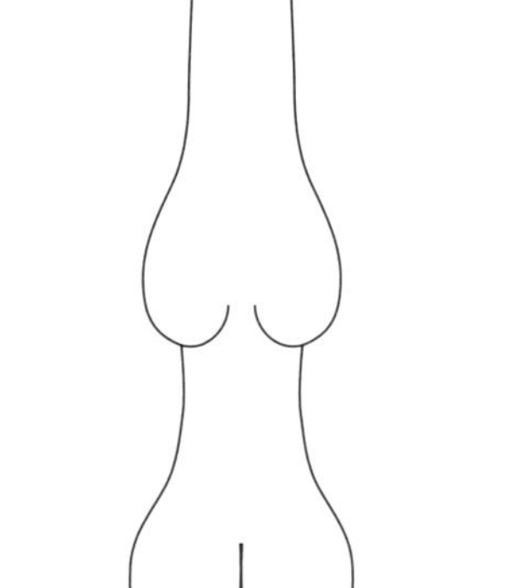

Thoracic duct receives lymph drained from all other areas of the body, and returns it to the blood circulation at the junction of the left jugular and left subclavian veins.

Cervical nodes

Inguinal nodes

The spleen is the largest mass of lymphatic tissue in the body, and has the role of filtration of ***blood*** (not lymph, like the other lymph nodes). In the spleen the blood, plus any potential pathogens, is exposed to lymphocytes which may then be triggered to produce appropriate antibodies. The spleen also phagocytoses bacterial and worn-out red blood cells, acts as a blood reservoir and, during fetal development, produces red blood cells.

Lymph nodes are located along the lymphatic vessels, usually in groups, some *superficial* (easily located during infection) and some *deep*. As lymph passes through these nodes it is filtered of foreign substances which are trapped within a network of fibres and then phagocytosed by *macrophages* or destroyed by products of *T-cells.* Lymph nodes also contain *B-lymphocytes,* which secrete antibodies, or may themselves leave the node and circulate to other parts of the body.

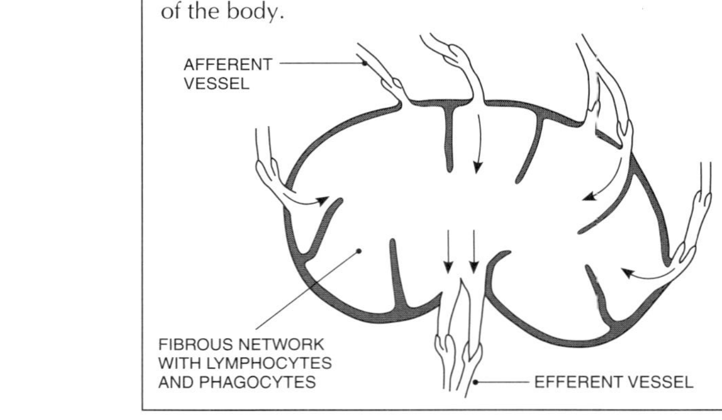

Relationship between lymphatic and blood vascular systems

Lymph returns to low pressure vessels of the blood vascular system.

Blood is forced into the systemic circulation at about 4.5 cm^3 per minute.

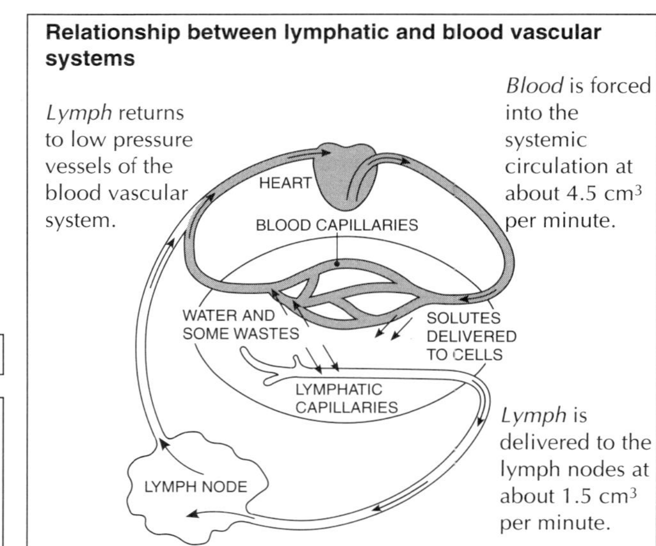

Lymph is delivered to the lymph nodes at about 1.5 cm^3 per minute.

At the lymph node, ***phagocytes*** remove toxins and pathogens, and ***B-lymphocytes*** secrete antibodies.

Control systems in biology

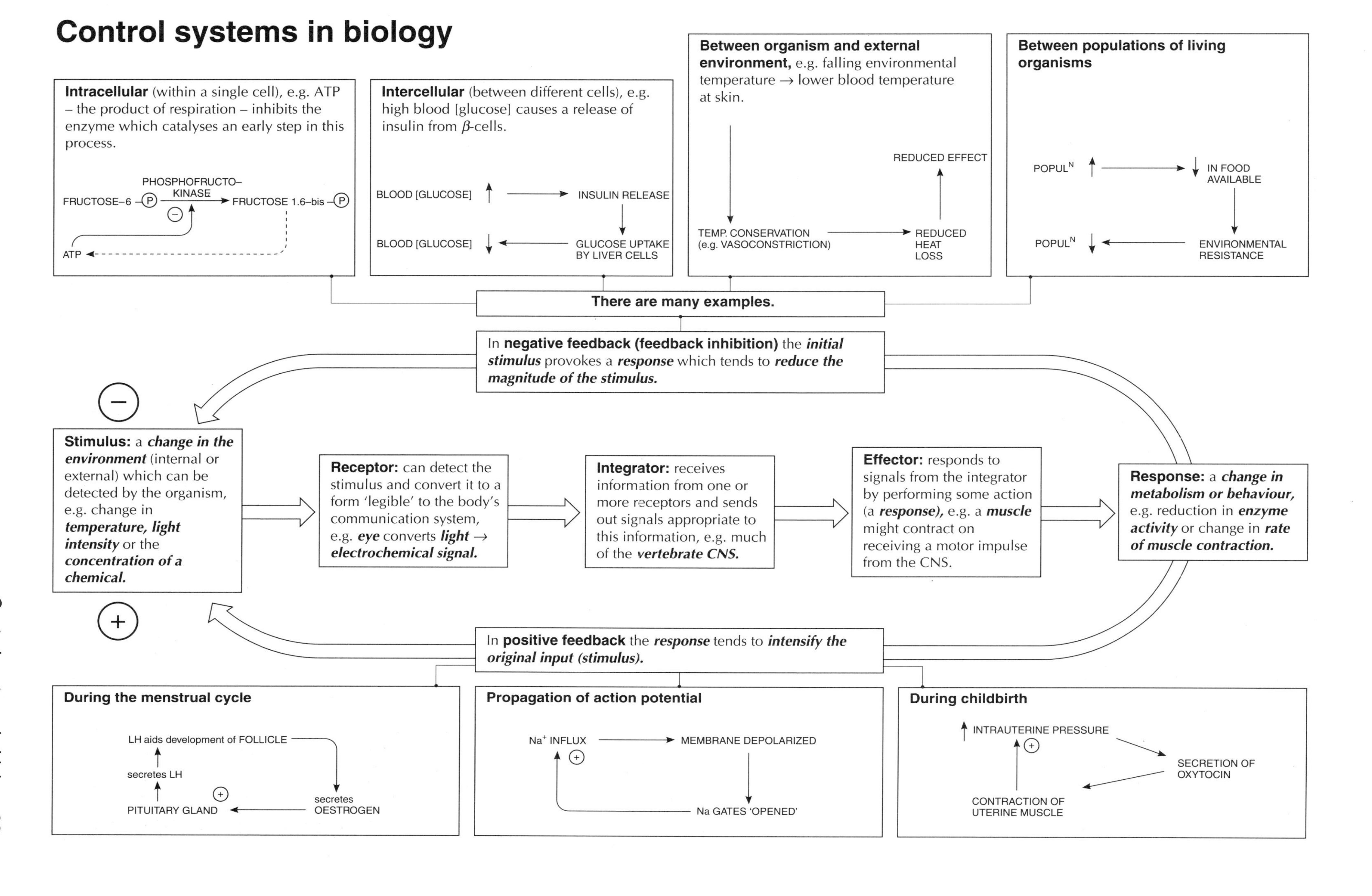

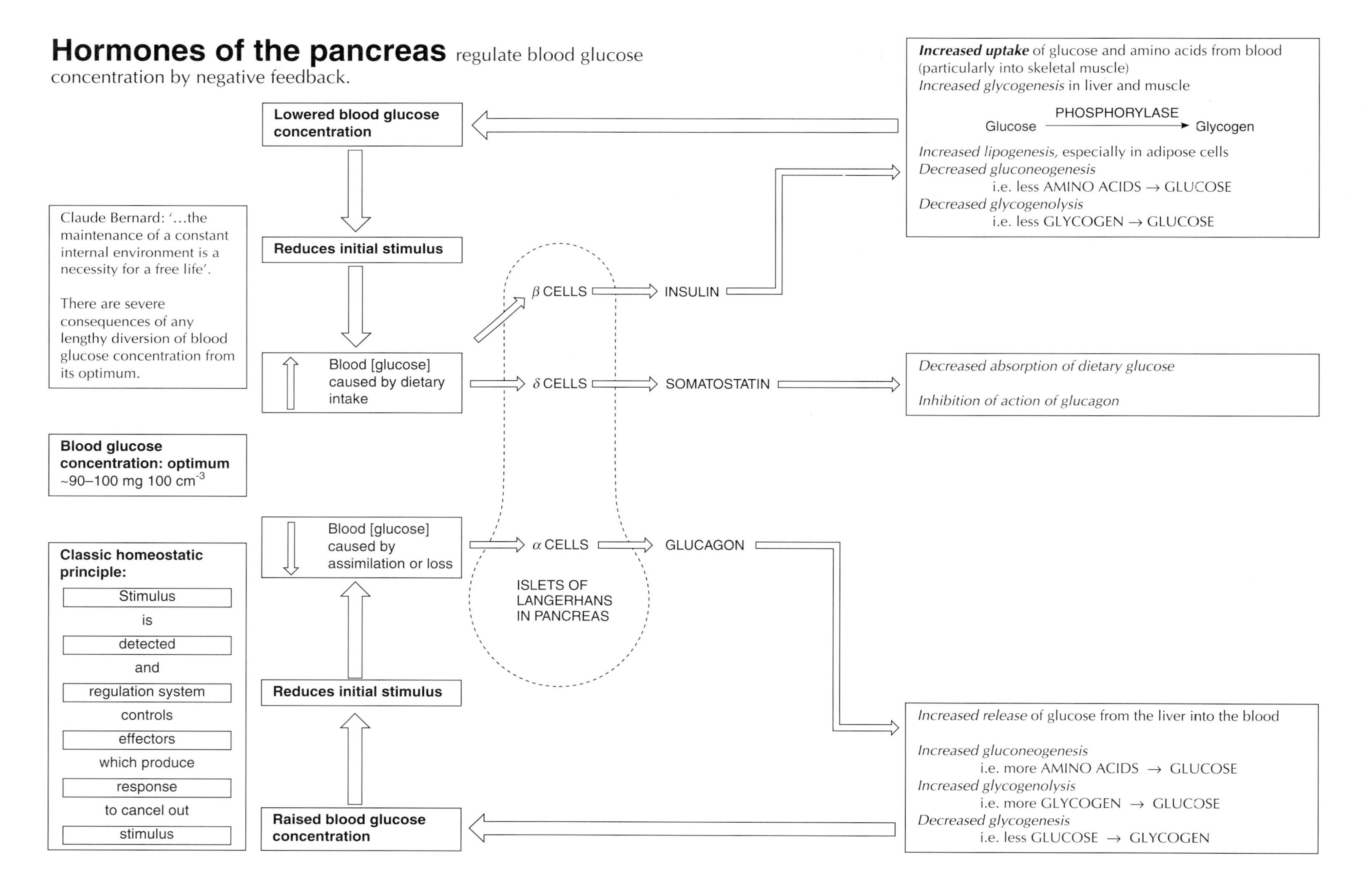

Hormones of the pancreas regulate blood glucose concentration by negative feedback.
Claude Bernard: '...the maintenance of a constant internal environment is a necessity for a free life'.
There are severe consequences of any lengthy diversion of blood glucose concentration from its optimum.
Blood glucose concentration: optimum ~90–100 mg 100 cm^{-3}
Classic homeostatic principle:
Stimulus
is
detected
and
regulation system
controls
effectors
which produce
response
to cancel out
stimulus
Lowered blood glucose concentration
Reduces initial stimulus
↑ Blood [glucose] caused by dietary intake
↓ Blood [glucose] caused by assimilation or loss
Reduces initial stimulus
Raised blood glucose concentration
β CELLS
δ CELLS
α CELLS
ISLETS OF LANGERHANS IN PANCREAS
INSULIN
SOMATOSTATIN
GLUCAGON
Increased uptake of glucose and amino acids from blood (particularly into skeletal muscle)
Increased glycogenesis in liver and muscle
Glucose —PHOSPHORYLASE→ Glycogen
Increased lipogenesis, especially in adipose cells
Decreased gluconeogenesis
i.e. less AMINO ACIDS → GLUCOSE
Decreased glycogenolysis
i.e. less GLYCOGEN → GLUCOSE
Decreased absorption of dietary glucose
Inhibition of action of glucagon
Increased release of glucose from the liver into the blood
Increased gluconeogenesis
i.e. more AMINO ACIDS → GLUCOSE
Increased glycogenolysis
i.e. more GLYCOGEN → GLUCOSE
Decreased glycogenesis
i.e. less GLUCOSE → GLYCOGEN

The urinary system

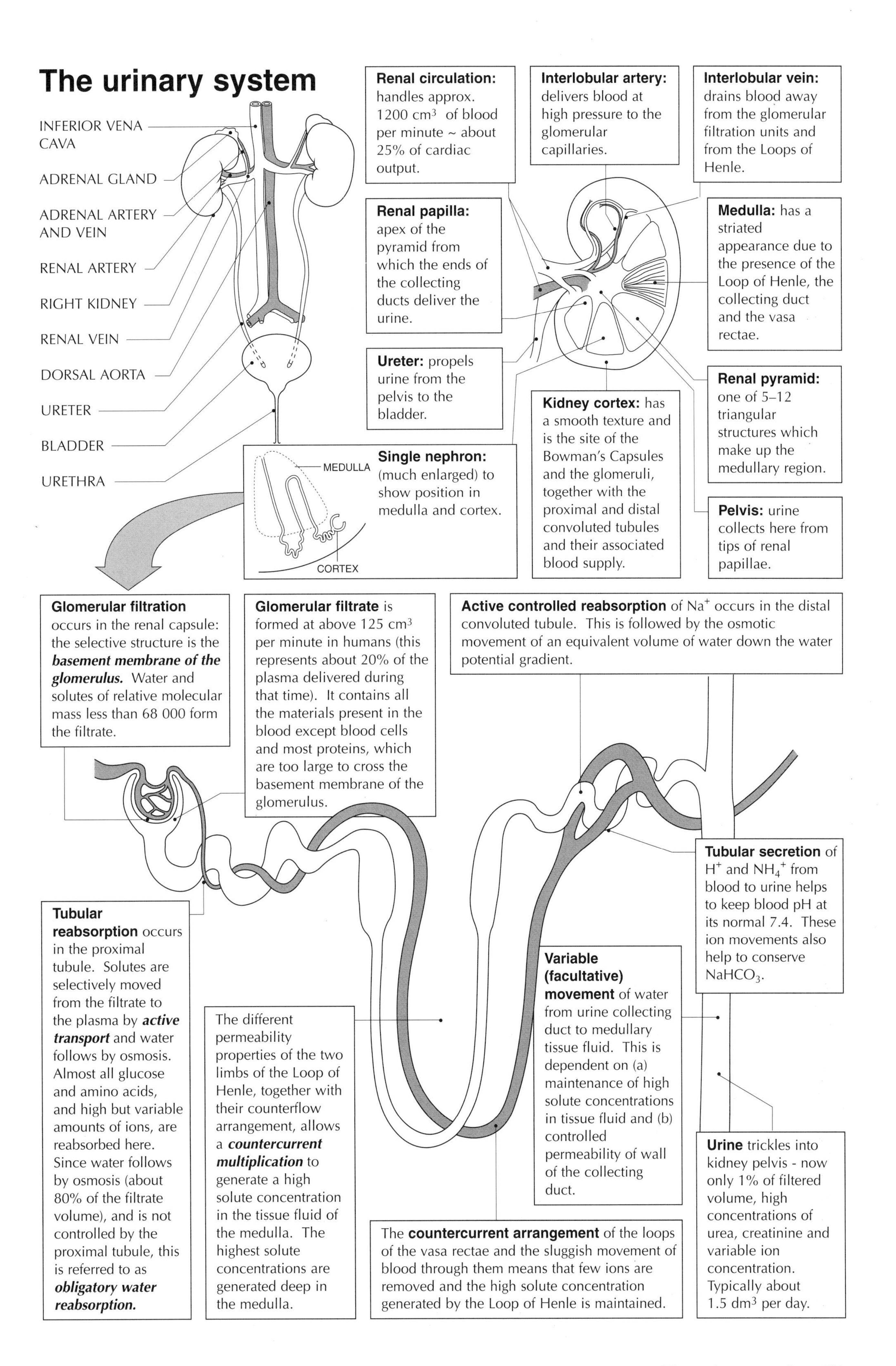

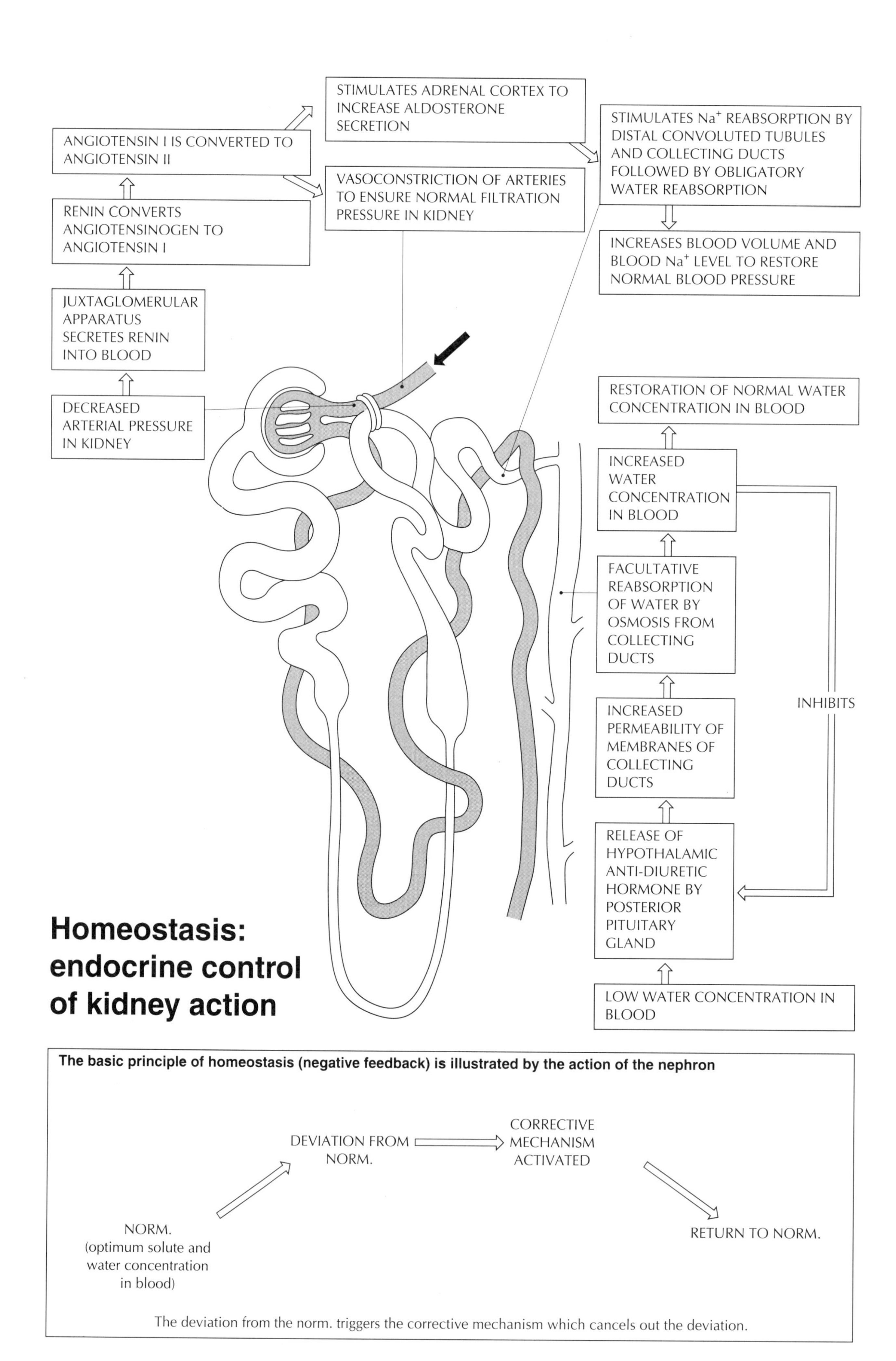

Homeostasis: endocrine control of kidney action

The basic principle of homeostasis (negative feedback) is illustrated by the action of the nephron

NORM. (optimum solute and water concentration in blood) → DEVIATION FROM NORM. → CORRECTIVE MECHANISM ACTIVATED → RETURN TO NORM.

The deviation from the norm. triggers the corrective mechanism which cancels out the deviation.

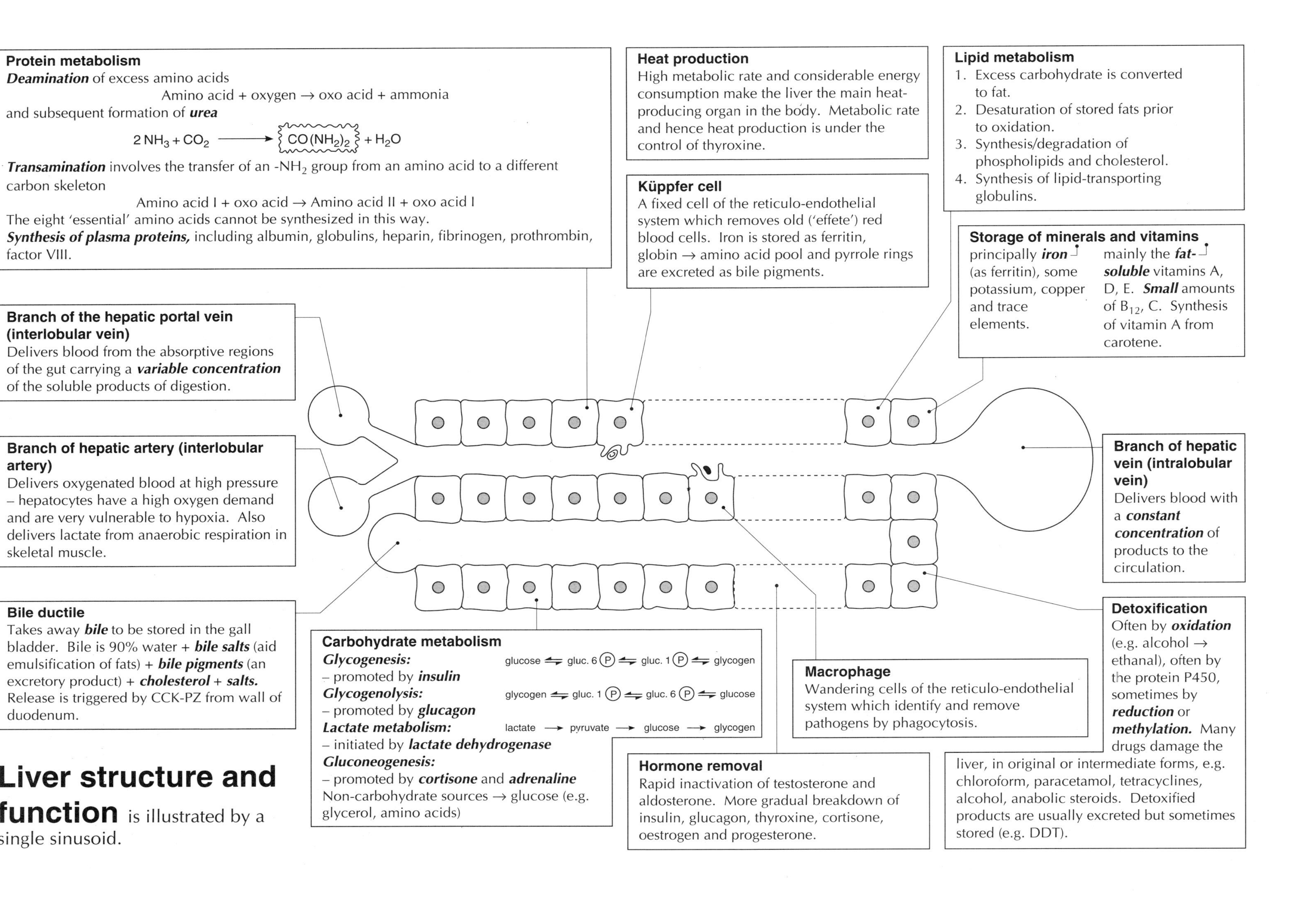
Protein metabolism
Deamination of excess amino acids
Amino acid + oxygen → oxo acid + ammonia
and subsequent formation of urea
$2NH_3 + CO_2 \longrightarrow CO(NH_2)_2 + H_2O$
Transamination involves the transfer of an $-NH_2$ group from an amino acid to a different carbon skeleton
Amino acid I + oxo acid → Amino acid II + oxo acid I
The eight 'essential' amino acids cannot be synthesized in this way.
Synthesis of plasma proteins, including albumin, globulins, heparin, fibrinogen, prothrombin, factor VIII.
Heat production
High metabolic rate and considerable energy consumption make the liver the main heat-producing organ in the body. Metabolic rate and hence heat production is under the control of thyroxine.
Lipid metabolism
1. Excess carbohydrate is converted to fat.
2. Desaturation of stored fats prior to oxidation.
3. Synthesis/degradation of phospholipids and cholesterol.
4. Synthesis of lipid-transporting globulins.
Küppfer cell
A fixed cell of the reticulo-endothelial system which removes old ('effete') red blood cells. Iron is stored as ferritin, globin → amino acid pool and pyrrole rings are excreted as bile pigments.
Storage of minerals and vitamins
principally iron (as ferritin), some potassium, copper and trace elements.
mainly the fat-soluble vitamins A, D, E. Small amounts of B_{12}, C. Synthesis of vitamin A from carotene.
Branch of the hepatic portal vein (interlobular vein)
Delivers blood from the absorptive regions of the gut carrying a variable concentration of the soluble products of digestion.
Branch of hepatic artery (interlobular artery)
Delivers oxygenated blood at high pressure – hepatocytes have a high oxygen demand and are very vulnerable to hypoxia. Also delivers lactate from anaerobic respiration in skeletal muscle.
Branch of hepatic vein (intralobular vein)
Delivers blood with a constant concentration of products to the circulation.
Bile ductile
Takes away bile to be stored in the gall bladder. Bile is 90% water + bile salts (aid emulsification of fats) + bile pigments (an excretory product) + cholesterol + salts.
Release is triggered by CCK-PZ from wall of duodenum.
Carbohydrate metabolism
Glycogenesis:
– promoted by insulin
glucose ⇌ gluc. 6 (P) ⇌ gluc. 1 (P) ⇌ glycogen
Glycogenolysis:
– promoted by glucagon
glycogen ⇌ gluc. 1 (P) ⇌ gluc. 6 (P) ⇌ glucose
Lactate metabolism:
– initiated by lactate dehydrogenase
lactate → pyruvate → glucose → glycogen
Gluconeogenesis:
– promoted by cortisone and adrenaline
Non-carbohydrate sources → glucose (e.g. glycerol, amino acids)
Macrophage
Wandering cells of the reticulo-endothelial system which identify and remove pathogens by phagocytosis.
Detoxification
Often by oxidation (e.g. alcohol → ethanal), often by the protein P450, sometimes by reduction or methylation. Many drugs damage the liver, in original or intermediate forms, e.g. chloroform, paracetamol, tetracyclines, alcohol, anabolic steroids. Detoxified products are usually excreted but sometimes stored (e.g. DDT).
Hormone removal
Rapid inactivation of testosterone and aldosterone. More gradual breakdown of insulin, glucagon, thyroxine, cortisone, oestrogen and progesterone.
Liver structure and function is illustrated by a single sinusoid.

Control of body temperature in mammals

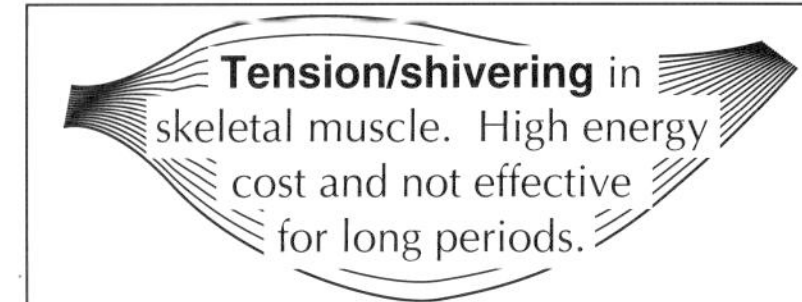

Tension/shivering in skeletal muscle. High energy cost and not effective for long periods.

EXTERNAL TEMPERATURE HIGH

Pilo-erector muscles relaxed: hair shafts 'flatten' and allow free circulation of air over hairs. Moving air is a good convector of heat.

EXTERNAL TEMPERATURE LOW

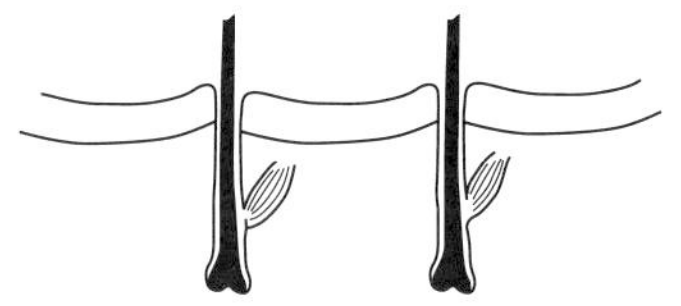

Pilo-erector muscles contracted: hair shafts perpendicular to skin surface. Trapped air is a poor conductor of heat so warm skin is insulated (similar effect by adding layers of clothing).

EXTERNAL TEMPERATURE HIGH

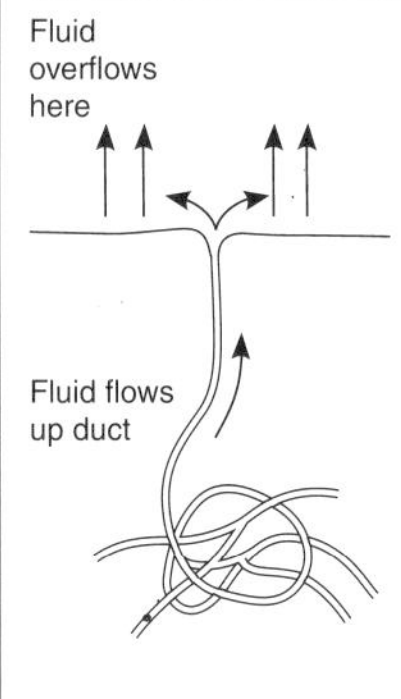

Sweat changed into vapour, taking latent heat of evaporation from the body to do this (about 2.5 kJ for each gram evaporated). Sweat glands extract larger volume of fluid from blood.

EXTERNAL TEMPERATURE LOW

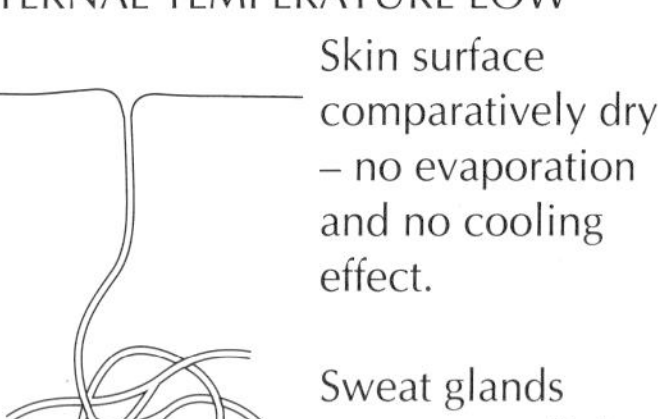

Skin surface comparatively dry – no evaporation and no cooling effect.

Sweat glands extract very little fluid from blood.

Non-shivering heat response triggers a general increase in metabolic rate. This is particularly noticeable in ***brown adipose tissue*** of newborns and animals that become acclimatized to cold. The ***liver*** of adults is also affected.

Changes in behaviour

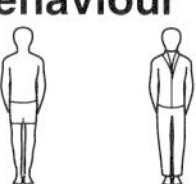

Dressing/undressing; moving in and out of shade; rest/activity cycles may all affect heat loss or production.

Hypothalamus contains the thermoregulatory centre which compares sensory input with a set point and initiates the appropriate motor responses. The set point may be raised by the action of pyrogens during pyrexia (fever).

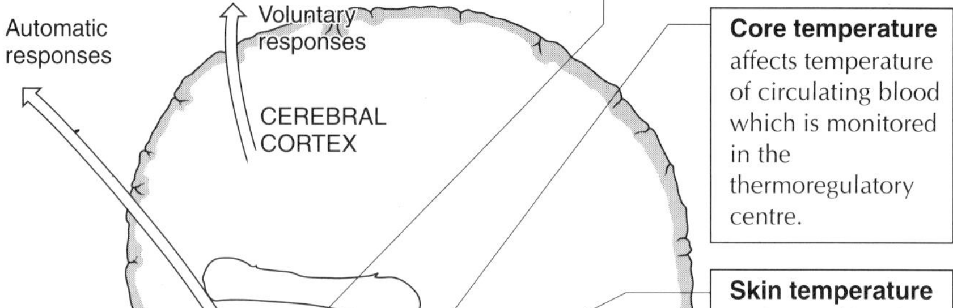

Core temperature affects temperature of circulating blood which is monitored in the thermoregulatory centre.

Skin temperature is detected by skin thermoreceptors which deliver sensory input to hypothalamus via cutaneous nerves.

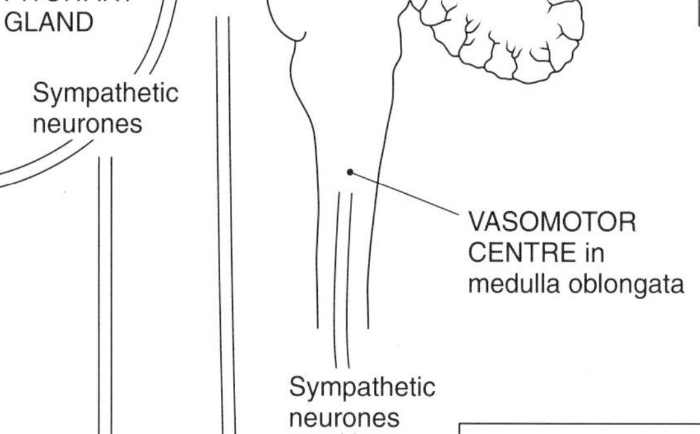

EXTERNAL TEMPERATURE HIGH

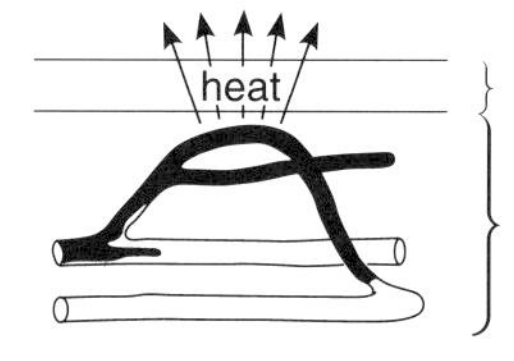

Vasodilation: sphincters/dilation of superficial arterioles allow blood close to surface. Heat lost by radiation – body cooled.

EXTERNAL TEMPERATURE LOW

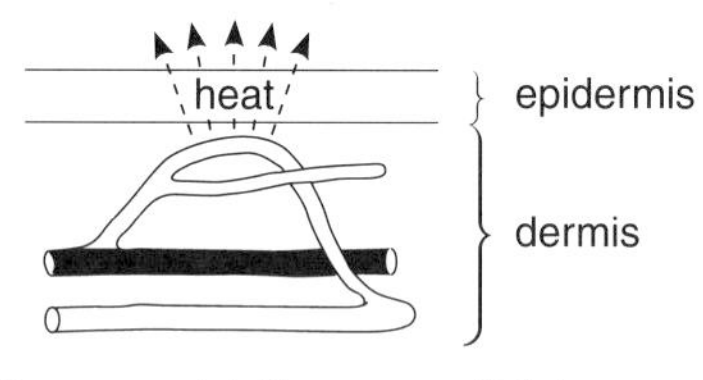

Vasoconstriction: superficial arterioles are constricted so that blood is shunted away from surface – heat is conserved.

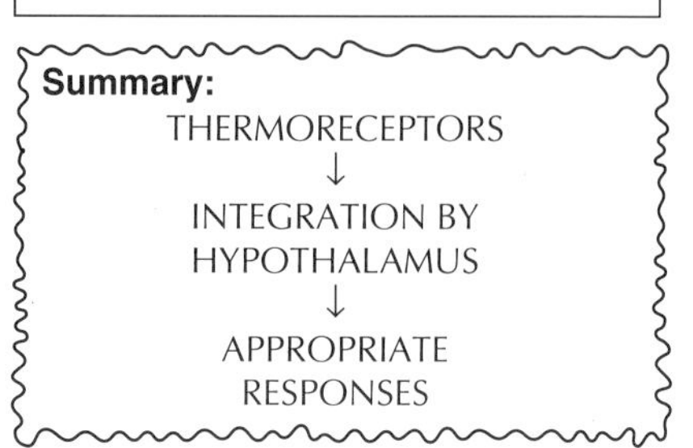

Summary:

THERMORECEPTORS
↓
INTEGRATION BY HYPOTHALAMUS
↓
APPROPRIATE RESPONSES

Ectotherms attempt to maintain body temperature by ***behavioural*** rather than ***physiological*** methods. These methods are less precise and so thermoregulation is more difficult.

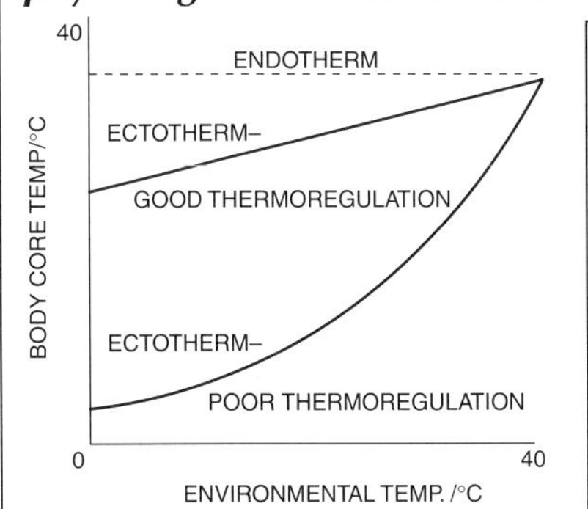

Reorientation of the body with respect to solar ***radiation*** can vary the surface area exposed to heating. A terrestrial ectotherm may gain heat rapidly by aligning itself at right angles to the Sun's rays but as its body temperature rises it may reduce the exposed surface by reorientating itself parallel to the Sun's rays.

Thermal gaping is used by some larger ectotherms such as alligators and crocodiles. The open mouth allows heat loss by ***evaporation*** from the moist mucous surfaces. Some tortoises have been observed to use a similar principle by spreading saliva over the neck and front legs which then acts as an evaporative surface.

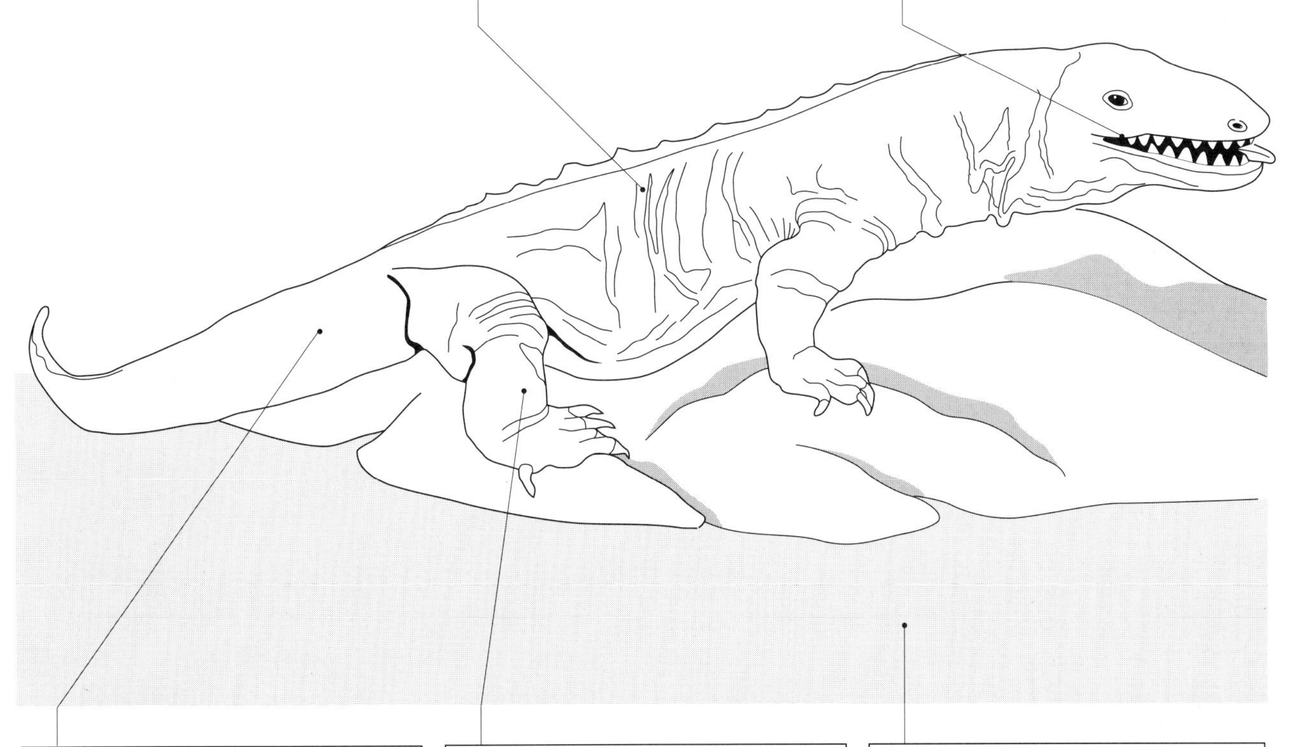

Colour changes of the skin may alter the ability of the body to absorb ***radiated*** heat energy. A dark-bodied individual will absorb heat more rapidly than a light-bodied one – thus some ectotherms begin the day with a dark body to facilitate 'warming-up' but then lighten the body as the environmental temperature rises.

Body raising is used by ectotherms to minimize heat gains by ***conduction*** from hot surfaces such as rocks and sand. The whole body may be lifted and the animal may reduce the area of contact to the absolute minimum by balancing on alternate diagonal pairs of feet.

Burrowing is a widely used behavioural device which enables ectotherms to avoid the greater temperature fluctuations on the surface of their habitat. The temperature in even a shallow burrow may only fluctuate by 5°C over a 24 h period whereas the surface temperature may range over 40°C during the same time. Amphibious and semi-aquatic reptiles such as alligators and crocodiles may return to water rather than burrow, since the high heat capacity of water means that its temperature is relatively constant.

The marine iguana and bradycardia

The marine iguana of the Galapagos Islands feeds by browsing on seaweed gathered from the sea around the rocky shores on which it lives. When basking on the rocks it normally maintains a body temperature of 37°C but during the time spent feeding in the sea it is exposed to environmental temperatures of 22–25°C. In order to avoid losing heat rapidly by ***conduction and convection*** the iguana reduces the flow of blood between its core tissues (at 37°C) and its skin (22°C) by slowing its heart rate (bradycardia).

Heat transfer between the body of an organism and its environment depends on the ***magnitude*** and ***direction*** of the ***thermal gradient*** (i.e. the temperature difference between the organism and its surroundings). Heat may be ***lost*** or ***gained*** by
conduction (heat transfer by physical contact)
convection (heat transfer to the air) and
radiation (heat transfer in the form of long-wave, infra-red electromagnetic waves)
but can only be ***lost*** by
evaporation (heat consumption during the conversion of water to water vapour).

The immune response

involves a wide range of cells and their products in defence against diseases.

NON-SPECIFIC IMMUNE RESPONSES: occur whenever ***any*** antigen passes the body's passive defences. These responses include:
phagocytosis
inflammation
release of interferon (anti-viral agent)
release of pyrogen (promotes fever)

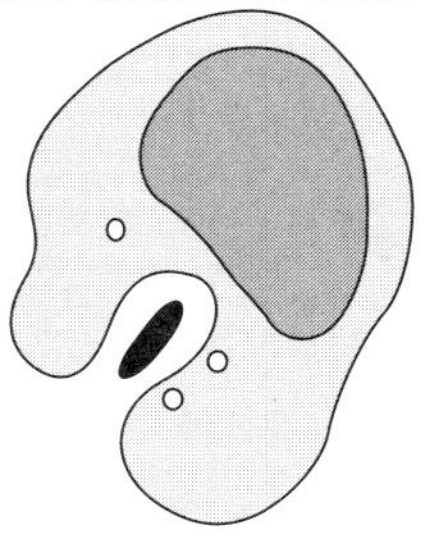

STIMULUS OR CHALLENGE IS AN ANTIGEN (molecule, usually protein or carbohydrate, recognized as 'non-self'). 'Self' cells are protected by the presence of ***MHC (major histocompatibility complex) protein*** on their surface.

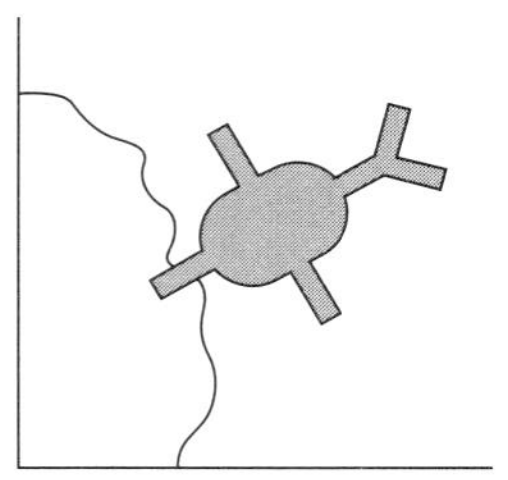

SPECIFIC IMMUNE RESPONSES: occur when a particular antigen passes the body's passive defences.

CELL-MEDIATED IMMUNE RESPONSE, in which a range of ***T-lymphocytes*** recognize, attack and destroy infected, mutant or 'foreign' cells.

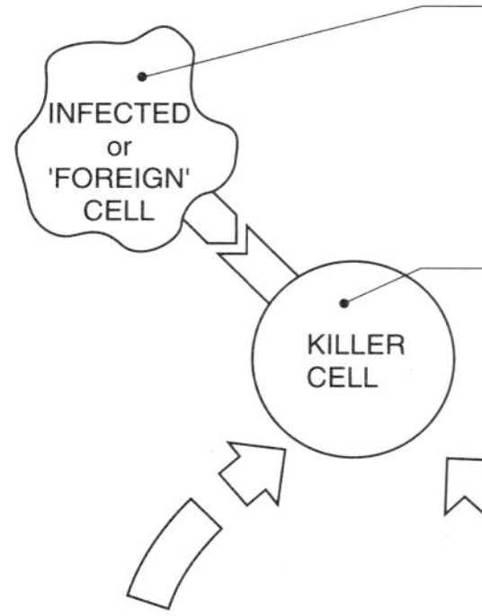

INFECTED or 'FOREIGN' CELL

Infected, mutant or 'foreign' cell presents surface antigen which is recognized as 'non-self' by macrophages or T cells.

KILLER CELL

Killer cell destroys target by release of ***perforin*** (punches holes in membrane) or ***nitric oxide*** (directly toxic).

Memory T cells are ready to respond rapidly to further challenges by the same antigen.

some clones reserved

Killer T cells ($T_{CYTOTOXIC}$ cells) attack cells infected by virus, bacteria or fungus. **Natural killer (N_K) cells** may attack tumour and other cancerous cells.

Macrophage may 'present' fragment of antigen ***(epitope)*** on surface.

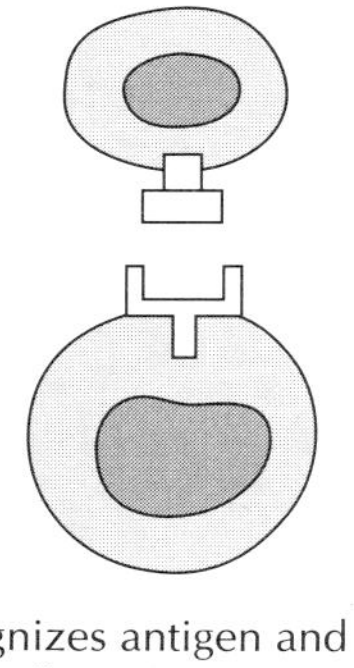

T-helper cell recognizes antigen and initiates activity of B and T cells.

Lymphokines/interleukins are chemical messengers.

+

T suppressor cells inhibit B and T cells once an immune response has successfully eliminated an antigen or mutant cell.

−

HUMORAL (ANTIBODY-MEDIATED) IMMUNE RESPONSE in which ***B-lymphocytes*** secrete ***antibodies*** to recognize and destroy bacteria, viruses, some fungi and protozoans.

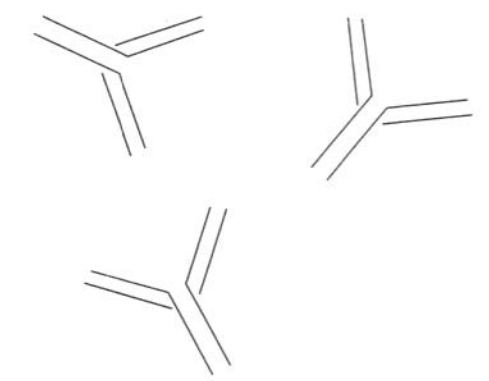

Antibody may directly remove or mask antigen ***(precipitation),*** mark antigen for destruction by macrophages ***(agglutination)*** or trigger lysis by a series of ***complement proteins.***

Plasma β-cells can synthesize and secrete up to 2000 antibody molecules per second.

Memory β-cells can develop into plasma cells if there are subsequent challenges by the same antigen.

many

a few

Cell division by MITOSIS produces a CLONE of identical cells

Virgin B-lymphocyte is able to rearrange its 'antibody-production' genes in an enormous number of combinations to deal with a specific antigen challenge.

Competent B-lymphocyte now has genotype to synthesize specific antibody.

Immune response II: antibodies and immunity

An antibody is a protein molecule synthesized by an animal in response to a specific antigen.

The basic structure of an antibody has the shape of the letter Y. Each molecule is composed of four polypeptide chains, two heavy and two light, all linked by disulphide bridges.

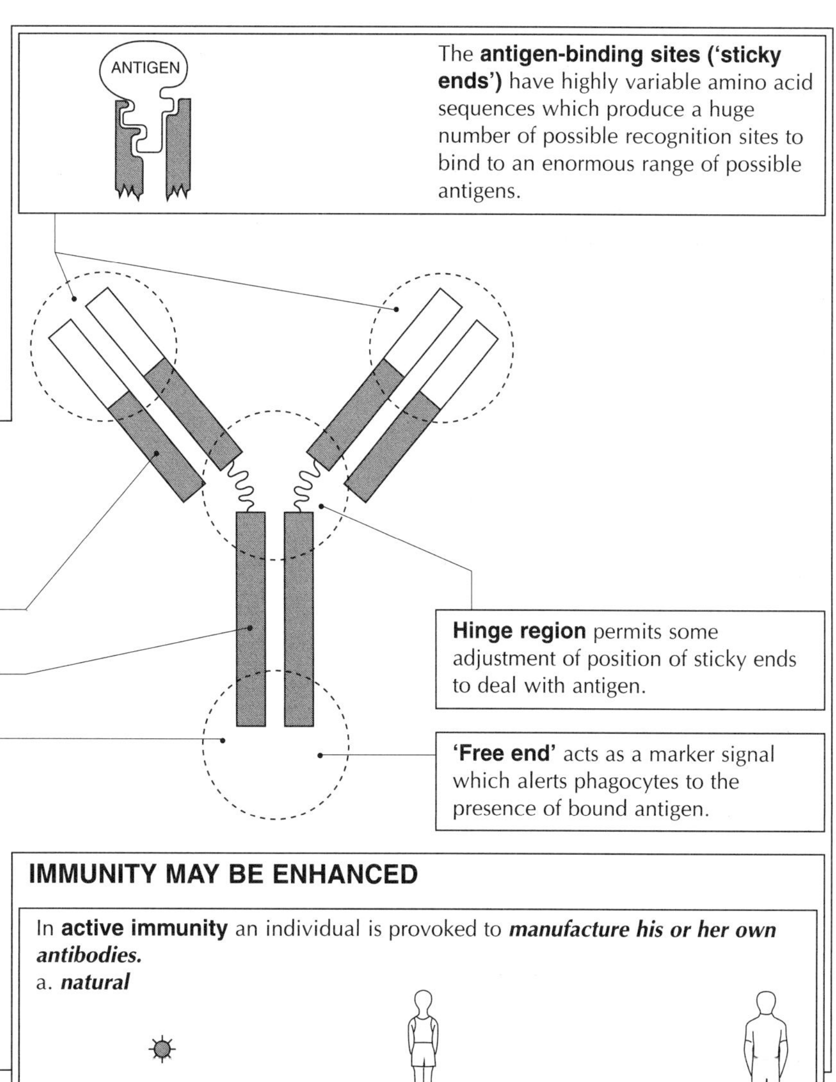

The constant regions determine the ***general class*** of the antibody:

IgG and IgM participate in the ***precipitation, agglutination*** and ***complement*** reactions.
IgA in tears, mucous secretions and saliva specifically binds to ***surface antigens on bacteria.***
IgD helps to ***activate lymphocytes.***
IgE is bound to mast cells and provokes ***allergies.***

IMMUNITY MAY BE ENHANCED

In **active immunity** an individual is provoked to ***manufacture his or her own antibodies.***

a. ***natural***

b. ***artificial***

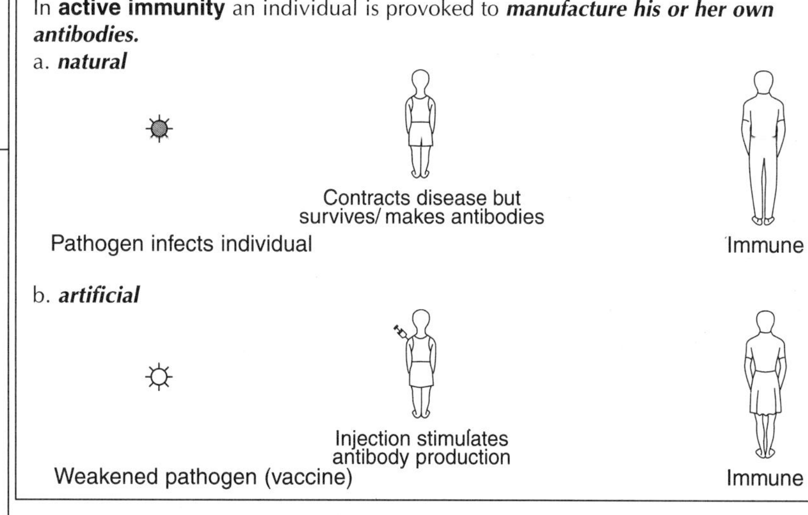

In **passive immunity** an individual is protected by a ***supply of pre-formed antibodies.***

a. ***natural***

b. ***artificial***

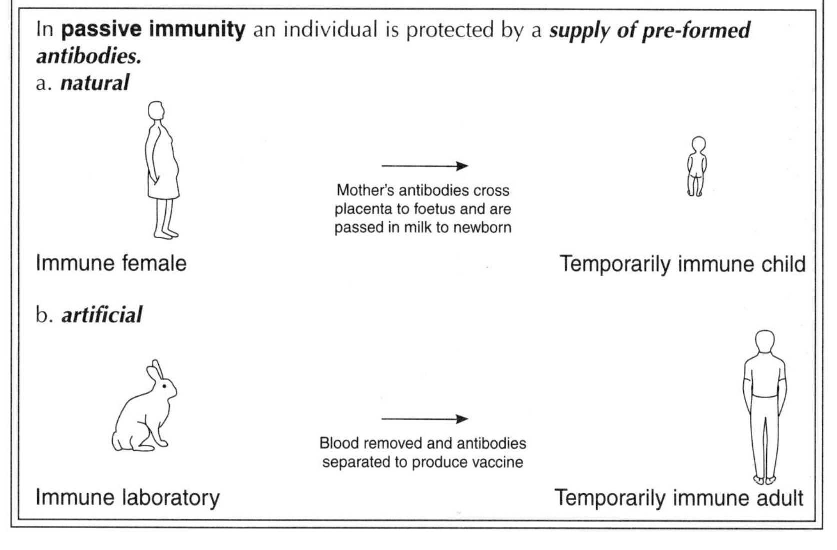

MEMORY CELLS SPEED UP IMMUNE RESPONSE

Secondary response: curve is steeper, peak is higher (commonly 10^3 x the primary response), lag period is negligible ***due to the presence of B-memory cells.*** Dominant antibody is IgG which is more stable and has a greater affinity for the antigen.

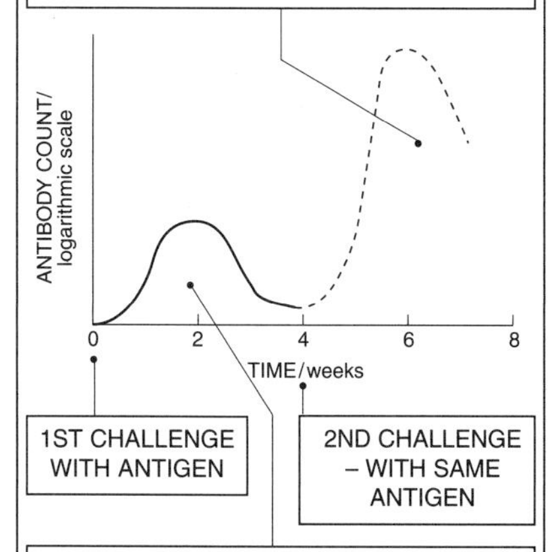

Primary response: typical lag period is 3 days with peak at 11–14 days. Dominant antibody molecule is IgM.

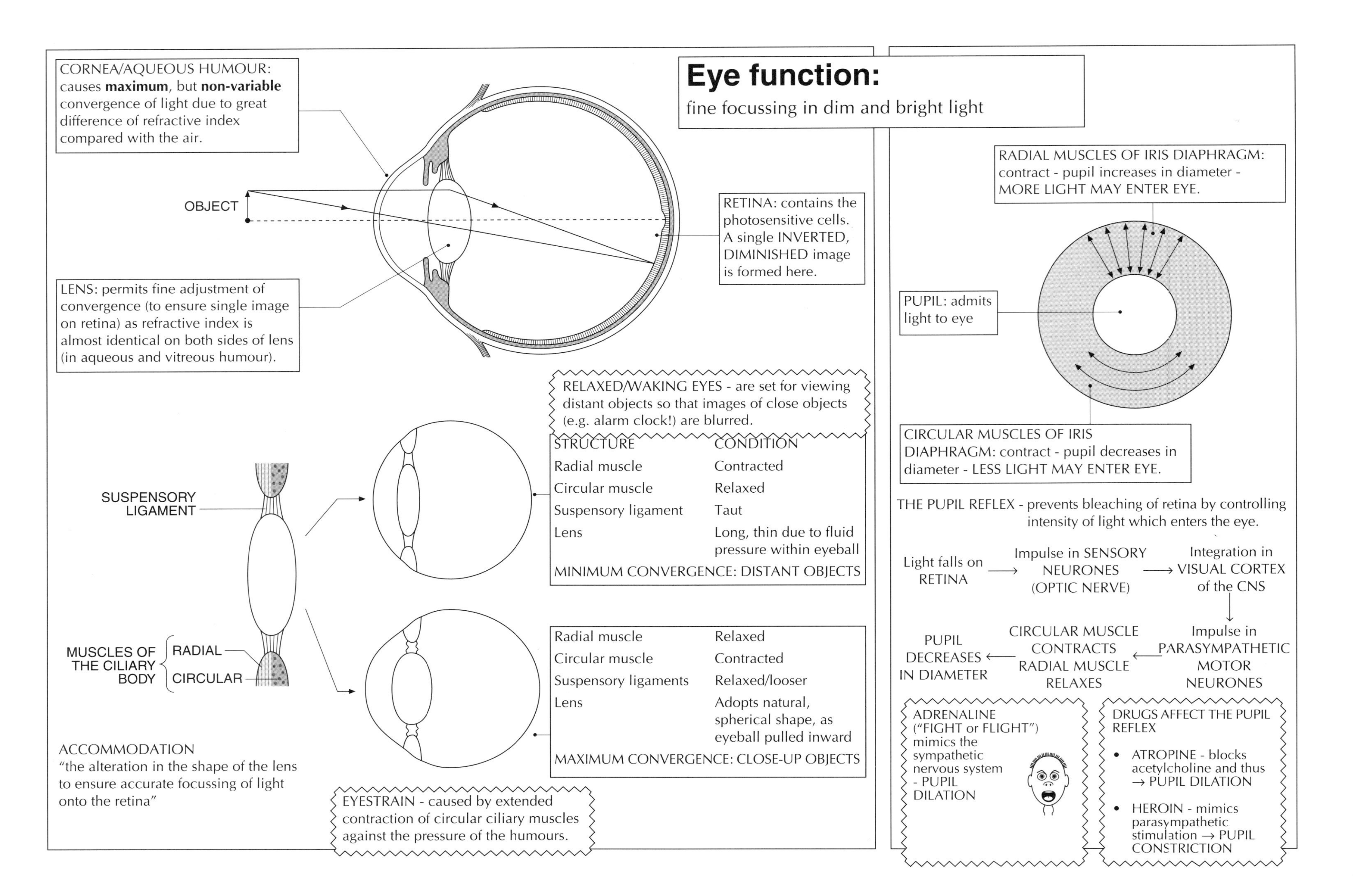
Eye function:
fine focussing in dim and bright light
CORNEA/AQUEOUS HUMOUR: causes maximum, but non-variable convergence of light due to great difference of refractive index compared with the air.
OBJECT
LENS: permits fine adjustment of convergence (to ensure single image on retina) as refractive index is almost identical on both sides of lens (in aqueous and vitreous humour).
RETINA: contains the photosensitive cells. A single INVERTED, DIMINISHED image is formed here.
SUSPENSORY LIGAMENT
MUSCLES OF THE CILIARY BODY
RADIAL
CIRCULAR
RELAXED/WAKING EYES - are set for viewing distant objects so that images of close objects (e.g. alarm clock!) are blurred.
STRUCTURE
CONDITION
Radial muscle
Contracted
Circular muscle
Relaxed
Suspensory ligament
Taut
Lens
Long, thin due to fluid pressure within eyeball
MINIMUM CONVERGENCE: DISTANT OBJECTS
Radial muscle
Relaxed
Circular muscle
Contracted
Suspensory ligaments
Relaxed/looser
Lens
Adopts natural, spherical shape, as eyeball pulled inward
MAXIMUM CONVERGENCE: CLOSE-UP OBJECTS
ACCOMMODATION
"the alteration in the shape of the lens to ensure accurate focussing of light onto the retina"
EYESTRAIN - caused by extended contraction of circular ciliary muscles against the pressure of the humours.
RADIAL MUSCLES OF IRIS DIAPHRAGM: contract - pupil increases in diameter - MORE LIGHT MAY ENTER EYE.
PUPIL: admits light to eye
CIRCULAR MUSCLES OF IRIS DIAPHRAGM: contract - pupil decreases in diameter - LESS LIGHT MAY ENTER EYE.
THE PUPIL REFLEX - prevents bleaching of retina by controlling intensity of light which enters the eye.
Light falls on RETINA
Impulse in SENSORY NEURONES (OPTIC NERVE)
Integration in VISUAL CORTEX of the CNS
Impulse in PARASYMPATHETIC MOTOR NEURONES
CIRCULAR MUSCLE CONTRACTS RADIAL MUSCLE RELAXES
PUPIL DECREASES IN DIAMETER
ADRENALINE ("FIGHT or FLIGHT") mimics the sympathetic nervous system - PUPIL DILATION
DRUGS AFFECT THE PUPIL REFLEX
• ATROPINE - blocks acetylcholine and thus → PUPIL DILATION
• HEROIN - mimics parasympathetic stimulation → PUPIL CONSTRICTION

Structure and function of the retina

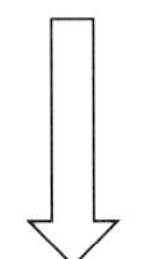

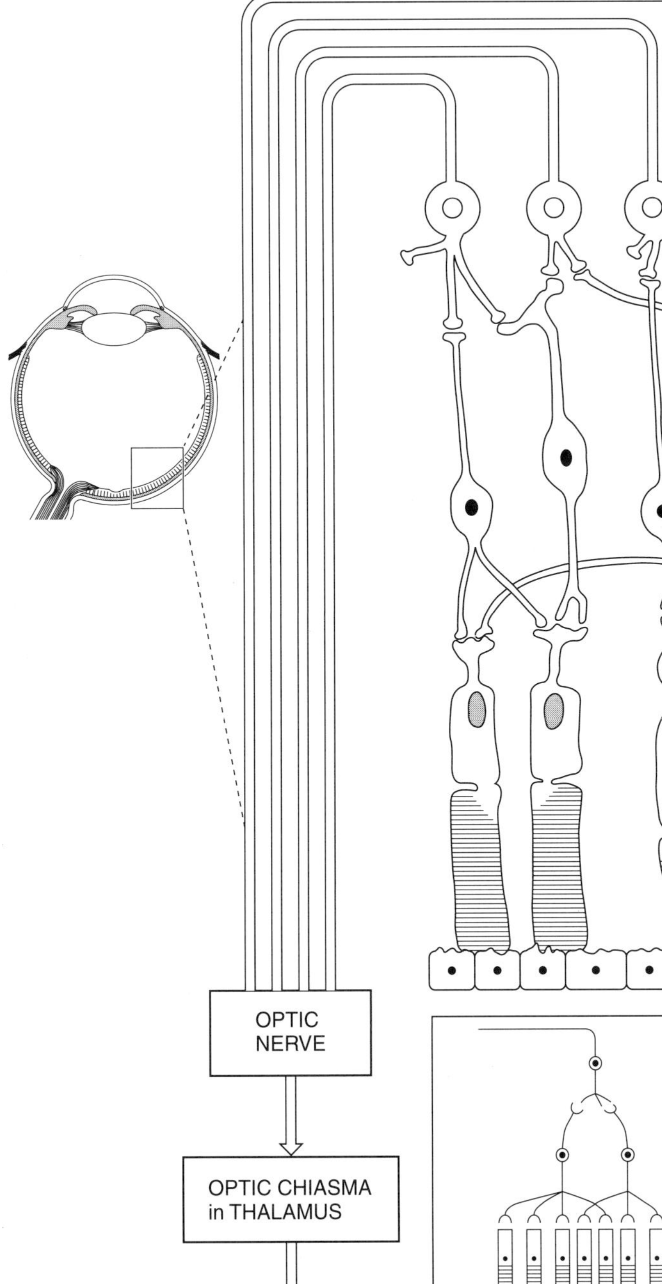

Ganglion cell: transmits action potential (depending on generator potential of photo-sensitive cells) along neurones of optic nerve.

Amacrine cell: interconnects sets of bipolar and ganglion cells – allows ***visual field pathways*** to influence one another and thus increase ***contrast and discrimination.***

Bipolar cell: these form an intermediate which connects photosensitive cells to ganglion cells – important in summation (see below).

Horizontal cell: interconnects rods and cones with bipolar cells – helps function of amacrine cells (see above).

Rod cell: sensitive to low levels of illumination but unable to discriminate between 'colours'.

Cone cell: contains pigment which is only sensitive to high levels of illumination but exists in different forms, so these cells can detect 'colours'.

Pigmented epithelium of retina

OPTIC NERVE → OPTIC CHIASMA in THALAMUS → PRIMARY VISUAL CORTEX → VISUAL ASSOCIATION CORTEX

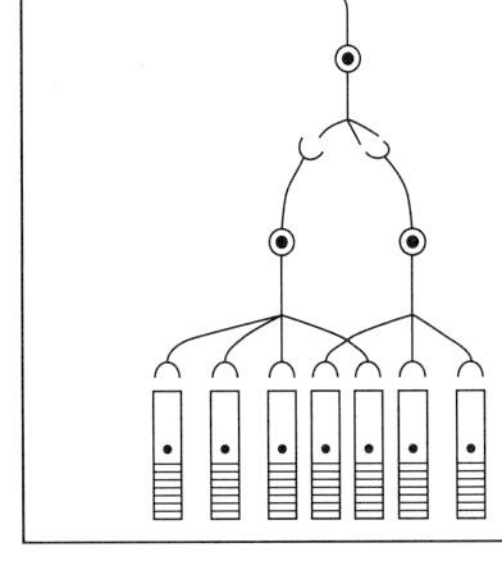

RODS, SUMMATION AND SENSITIVITY

The responses of many rods may be 'summed' by the anatomical arrangement of the bipolar cells, which may synapse with several rods but only a single ganglion cell. This ***synaptic convergence*** permits great ***visual sensitivity,*** and rods are thus of great value for ***night vision.*** Since rods are more abundant away from the fovea, objects are often seen more clearly at night by not looking directly at them.

CONES AND VISUAL ACUITY

Each cone is connected, via a bipolar cell, to a single ganglion cell. Since cones are packed closely together, especially at the fovea, these cells are able to discriminate between light stimuli which arrive in close proximity – the retina is able to resolve two light sources falling on cones separated only by a single other cone.
The ***absence of synaptic convergence*** offers ***acuity*** but ***poor sensitivity.***

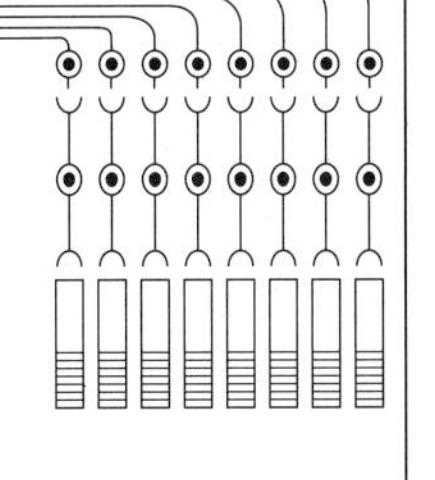

Endocrine control depends upon chemical messengers secreted from cells and binding to specific hormone receptors

A hormone is a chemical produced in one part of an organism which is transported throughout the organism and produces a specific response in target cells.

Hormones fall into ***three broad chemical categories***

1. ***Proteins*** (e.g. insulin) and ***peptides*** (e.g. oxytocin).

OXYTOCIN

S ——— S

CYS – TYR – ILE – GLN – ASN – CYS – PRO – LEU – GLY — NH_2

2. ***Amines*** (e.g. adrenaline) are derivatives of amino acids.

HO, HO (ring) — C(H)(OH) — CH_2 — NH — CH_3

3. ***Steroids*** (e.g. testosterone) derived from cholesterol.

CH_3 OH CH_3 O

Hormone-secreting cells may be

a. ***Endocrine*** – secrete hormones into the bloodstream
e.g. ***adrenal medulla/adrenaline***

b. ***Paracrine*** – secrete hormones that affect adjacent cells
e.g. ***gastric mucosa/gastrin/gastric pits***

c. ***Autocrine*** – regulate their own activity by the secretion of hormones
e.g. ***interstitial cells of testis/testosterone***

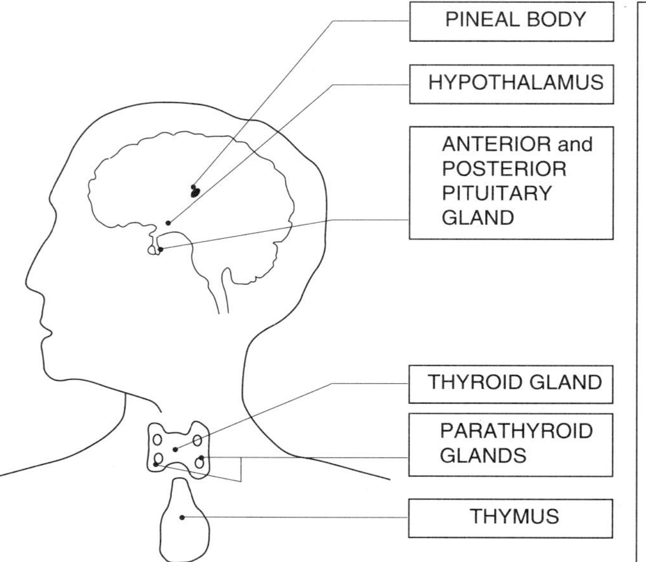

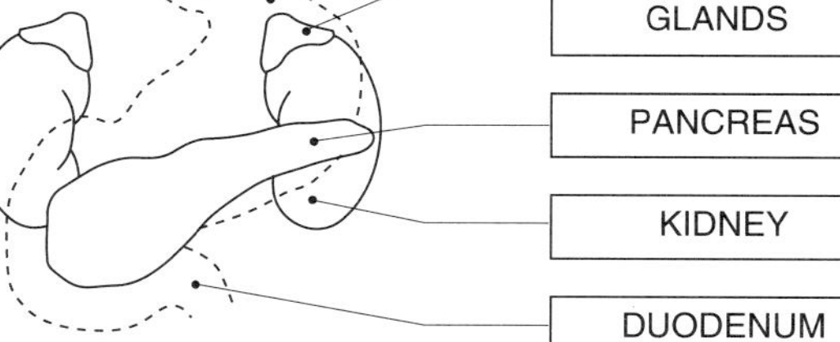
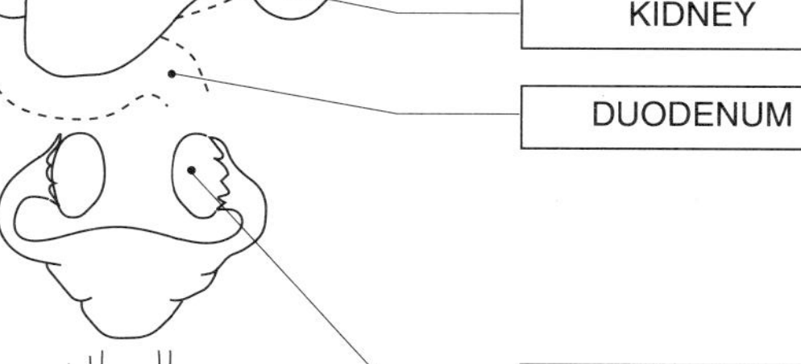

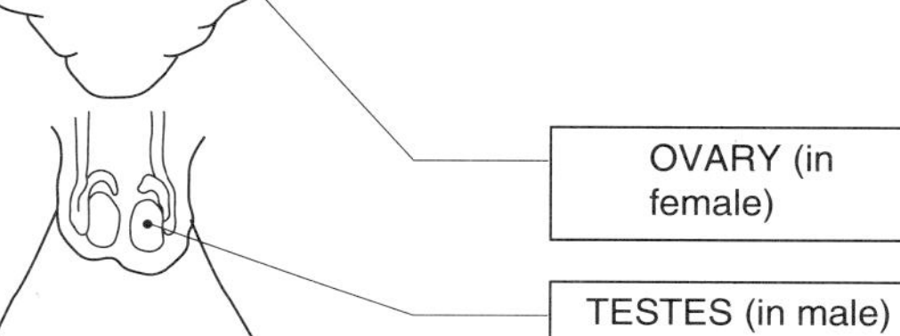

General principles of hormone activity

1. The specificity of hormone action depends upon ***target cells*** – these possess ***receptor molecules*** made of protein in their membranes, cytoplasm or nucleus and ***transduction machinery*** which can be activated by the ***hormone–receptor complex.***

2. Each target cell in the body is regulated only by those hormones to which it has receptor molecules, and not by others.

3. Different cells may respond in different ways to the same hormone – the transduction machinery of different target cell types 'reads' the hormone signal in different ways.

4. Some hormones, such as those which maintain solute concentrations in body fluids (e.g. insulin and calcitonin), are present for much of the time, whereas others, such as adrenaline (stress response) and gastrin (enzyme release), are secreted only when needed.

5. Blood hormone concentrations are usually governed by ***negative feedback control*** – a reduction in concentration stimulates additional secretion and an increase in concentration inhibits further secretion.

6. Once hormones bind to receptor molecules they are usually degraded rapidly. This 'rapid recovery system' means that target cells can be sensitive to changing levels of the hormones that regulate their activities.

Endocrine secretions in humans

ENDOCRINE GLAND AND HORMONE	TARGET	PRINCIPAL ACTION
Posterior pituitary gland secretes ***Oxytocin***	Uterus	Stimulates contraction during childbirth
	Mammary gland	Stimulates ejection of milk into ducts
Anti-diuretic hormone	Collecting ducts of kidney	Reabsorption of water from urine
Anterior pituitary gland secretes many hormones including ***trophins***		Trophins stimulate other endocrine organs
Thyroid gland secretes ***thyroxine*** and ***triiodothyronine***	Many tissues	Stimulation of metabolic rate
Parathyroid glands secrete ***parathormone***	Bone	Stimulates release of Ca^{2+}
	Kidneys	Stimulates Ca^{2+} reabsorption
	Gut	Activates vitamin D
Kidney secretes ***erythropoietin***	Bone marrow	Stimulates synthesis and maturation of erythrocytes
Islets of Langerhans in ***pancreas*** secrete ***insulin*** and ***glucagon***	General	Lowers blood glucose concentration
	Liver, adipose tissue	Raises blood glucose concentration
Testis secretes ***Testosterone***	General	Development of male secondary sexual characteristics
	Seminiferous tubules	Promotes spermatogenesis
Inhibin	Anterior pituitary gland	Controls sperm production by inhibition of FSH secretion

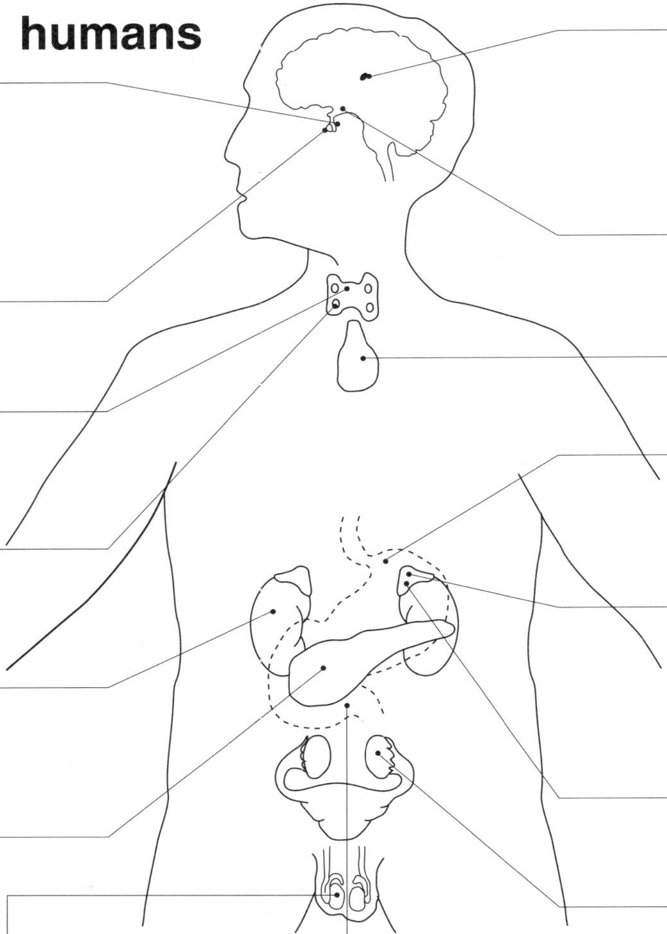

ENDOCRINE GLAND AND HORMONE	TARGET	PRINCIPAL ACTION
Duodenal mucosa secretes ***secretin***	Pancreas	Stimulates release of $NaHCO_3$ into pancreatic juice
Cholecystokinin	Pancreas	Stimulates release of digestive enzymes
	Gall bladder	Stimulates contraction and emptying

ENDOCRINE GLAND AND HORMONE	TARGET	PRINCIPAL ACTION
Pineal gland secretes ***melatonin***	Gonads	Seasonal control of reproductive activity in some mammals
	Melanocytes	Changes in pigmentation
Hypothalamus secretes ***releasing*** and ***release-inhibiting factors***	Anterior lobe of pituitary gland (adeno-hypophysis)	Control secretion of specific hormones, including the trophins
Thymus gland secretes ***thymosin*** and ***thymopoietin***	Lymphoid tissue	Maturation of T cells during cell-mediated immune response
Stomach secretes ***gastrin***	Gastric glands of stomach	Stimulates secretion of pepsinogen
Adrenal cortex secretes ***aldosterone*** and ***cortisol***	Kidney tubules	Increase blood levels of sodium and water and decrease potassium level
	General-many tissues	Promote resistance to stress; counter inflammatory responses
Adrenal medulla secretes ***adrenaline***	General	Mobilizes glucose, increases blood flow from heart among many responses to danger or stress
Ovary secretes ***Oestrogen***	General	Development of female secondary sexual characteristics
	Uterus	Repair of endometrium following menstruation
Progesterone	Uterus	Preparation for implantation
	Breast	Preparation of mammary glands for lactation
Relaxin	Pubic symphysis	Allows expansion of pelvis
	Cervix	Dilation at childbirth

Motor (efferent) neurone: the dendrites (antennae), axon (cable), synaptic buttons (contacts) are serviced and maintained by the cell body.

Nissl granules (or 'chromatophilic substance' because they take up stain readily) represent a highly ordered ***rough endoplasmic reticulum.*** Proteins made here, and passed into the neuronal processes (especially the axon), include structural proteins of the neurone membrane, transport proteins such as the Na^+/K^+ pump and enzymes involved in neurotransmitter synthesis.

Dendrites are extensions of the cell body containing all typical cell body organelles. They provide a large surface area to receive information which they then pass on towards the cell body. The plasma has a high density of ***chemically gated ion channels,*** important in impulse transmission.

Cell body contains a well-developed nucleus and nucleolus and many organelles such as lysosomes and mitochondria. Many neurones also contain yellowish-brown granules of ***lipofuscin pigment,*** which may be a by-product of lysosomal activity and which increases in concentration as the neurone ages. There is ***no mitotic apparatus*** (centriole/spindle) in neurones more than six months old, which means that damaged neurones can never be replaced (although they may regenerate – see below).

Neurofibrils are formed from microtubules and microfilaments. They offer support to the cell body and may be involved in the transport of materials throughout the neurone.

Axon hillock is the point on the neuronal membrane at which a ***threshold stimulus*** may lead to the initiation of an ***action potential.***

Axon collateral is a side branch of the axon which means that one cell may direct impulses to more than one effector.

Nodes of Ranvier are unmyelinated segments of the neurone. Since these are uninsulated, ion movements may take place which effectively lead to action potentials 'leaping' from one node to another during ***saltatory conduction.***

Schwann cell is a glial cell which encircles the axon. When the two 'ends' of the Schwann cell meet, overlapping occurs which pushes the nucleus and cytoplasm to the outside layer.

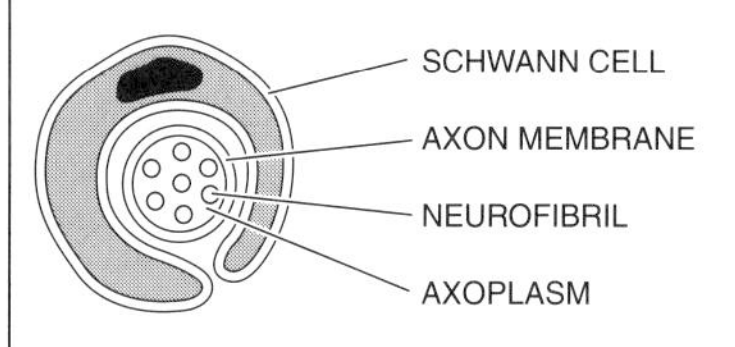

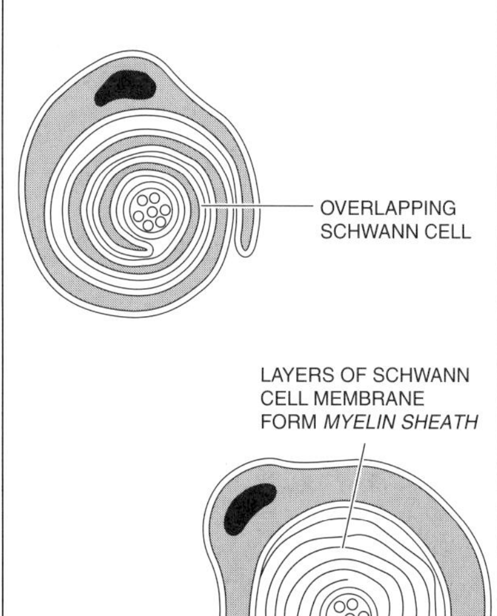

Axon is the communication route between the cell body and the axon terminals. There are two intracellular transport systems: ***axoplasmic flow*** is slow, unidirectional protoplasmic streaming which supplies new axoplasm for new or regenerating neurones; ***axonal transport*** is faster, bi-directional and via microtubules and microfilaments. Axonal transport returns materials to the cell body for degradation/recycling *but* is the route taken by the ***herpes virus*** and the ***rabies virus*** to the cell body, where they multiply and cause their damage. The toxin produced by the ***tetanus bacterium*** uses the same route to reach the central nervous system.

Axon teminal

Neurilemma is found only around axons of the peripheral nervous system, i.e. typical sensory and motor neurones. The neurilemma plays a part in the regeneration of damaged nerves by forming a tubular sheath around the damaged area within which regeneration may occur.

Myelin sheath is composed of 20–30 layers of Schwann cell membrane. The high phospholipid content of the sheath offers electrical insulation → ***saltatory impulse conduction.*** Not complete until late childhood so infants often have slow responses/poor co-ordination. Some axons are not myelinated but are enclosed in Schwann cell cytoplasm.

Synaptic end bulb or ***synaptic button*** is important in nerve impulse conduction from one neurone to another or from a neurone to an effector. They contain membrane-enclosed sacs (***synaptic vesicles***) which store ***neurotransmitters*** prior to release and diffusion to the post-synaptic membrane.

Spinal cord and reflex action

Reflex actions are rapid responses to internal or external stimuli which allow the body to maintain homeostasis. They may be ***somatic*** or ***autonomic*** but all involve the sequence

RECEPTOR → SENSORY NEURONE → CNS → MOTOR NEURONE → EFFECTOR

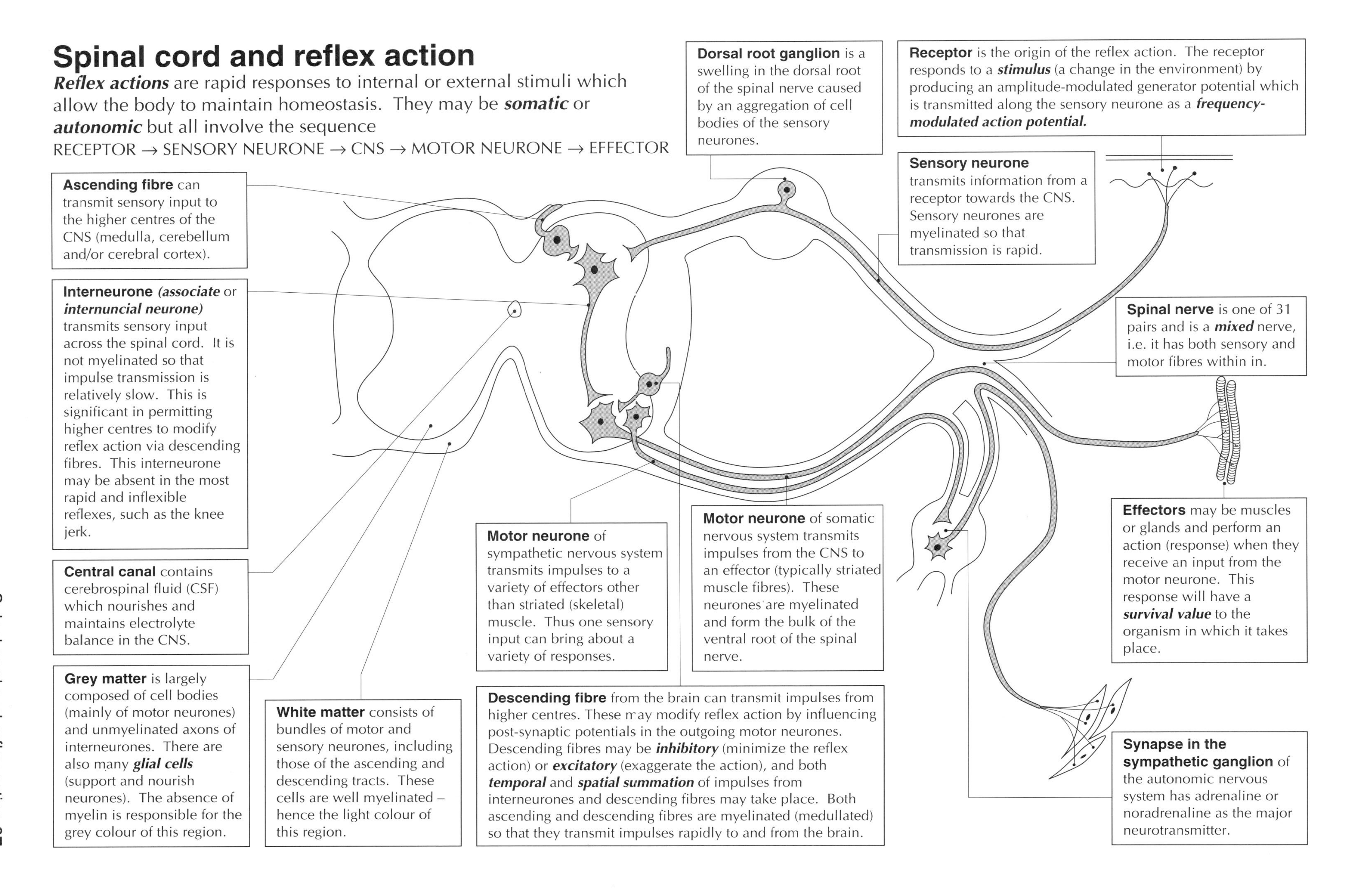

Resting potential and action potential:

The **resting potential** of -70mV is largely the result of a potassium ion (K^+) equilibrium.

SODIUM CHANNELS are pore proteins which permit VERY LIMITED DIFFUSION of Na+ from the extracellular fluid to the cytoplasm of the nerve cell.

ATP-DEPENDENT Na^+-K^+ PUMP transports ions across the plasamembrane of the nerve cell. Three Na^+ ions are moved from the inside to the outside of the cell for every two K^+ which are moved in the opposite direction, thus establishing ION CONCENTRATION GRADIENTS across the membrane.

POTASSIUM CHANNELS are pore proteins which permit K^+ ions to diffuse from the nerve cell cytoplasm to the extracellular fluid along a concentration gradient. These channels ensure that membrane permeability to K^+ is about 100x greater than that to Na^+ ions. K^+ movement continues until concentration gradient is 'balanced' by electrochemical gradient.

THE CYTOPLASM OF THE NERVE CELL contains many negative ions, mostly protein anions. The nerve cell membrane is almost impermeable to these ions.

NET Na^+ ION MOVEMENTS generate a lesser INSIDE - OUTSIDE K^+ ION GRADIENT

NET Na^+ ION MOVEMENTS generate a high OUTSIDE - INSIDE Na^+ ION GRADIENT

As a result of

a. Na^+/K^+ pump
b. K^+ diffusion outwards
c. negative ions in the nerve cell cytoplasm

the plasamembrane of the nerve cell becomes POLARISED: the inside is about -70mV with respect to the outside. This represents the RESTING POTENTIAL of the nerve cell.

−70mV

Return to resting potential: although the action potential involves Na^+ and K^+ movements, the changes in absolute ion concentrations are very small (probably no more than 1 in 10^7). Many action potentials could be transmitted before concentration gradients are significantly changed – the sodium–potassium pump can quickly restore resting ion concentrations to 'pre-impulse' levels.

An **action potential** is the depolarization–repolarization cycle at the neurone membrane following the application of a threshold stimulus. The depolarization is about 110 mV resulting from ***an inward flow of Na+ ions.*** Since the depolarization– repolarization depend upon ion concentration gradients and upon time of ion channel opening, both of which are effectively fixed, ***all action potentials are of the same size.*** Thus a nerve cell obeys the ***all-or-nothing principle:*** if a stimulus is strong enough to generate an action potential, the impulse is conducted along the entire neurone ***at a constant and maximum strength*** for the existing conditions.

MEMBRANE POTENTIAL/mV

TIME / ms

Depolarization: the voltage-gated sodium channels open so that Na^+ ions can move ***into the axon.***

a. down a ***Na^+ concentration gradient***

b. down an ***outside–inside electrical gradient***

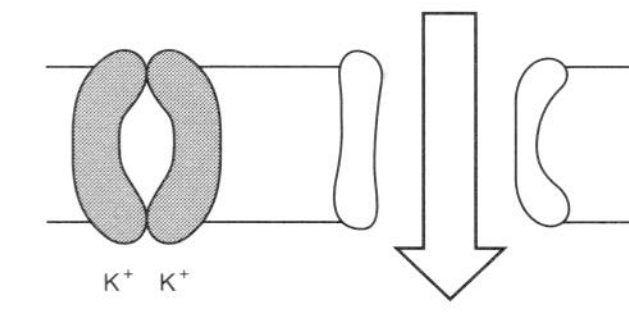

The inward movement of Na^+ ions during depolarization is an example of a ***positive feedback system.*** As Na^+ ions continue to move inward depolarization increases, which opens more sodium channels so more Na^+ ions enter causing more depolarization and so on.

Repolarization: sodium channels are closed but potassium channels are open so that K^+ ions are able to move ***out of the axon.***

a. down a K^+ ***concentration gradient***

b. down an ***electrochemical gradient***

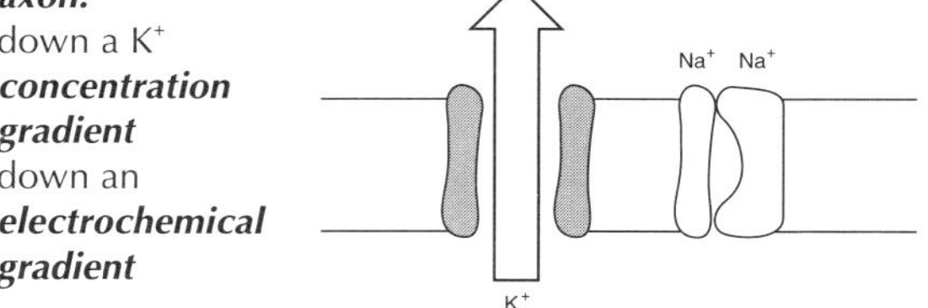

Threshold value: any stimulus strong enough to initiate an impulse is called a ***threshold*** or ***liminal stimulus.*** The stimulus begins the depolarization of the neuronal membrane – once a sufficient number of voltage-gated sodium channels is opened, positive feedback will ensure a complete depolarization. Any stimulus weaker than a threshold stimulus is called a ***sub-threshold*** or ***subliminal stimulus.*** Such a stimulus is incapable of initiating an action potential, but a series of such stimuli ***may*** exert a cumulative effect which may be sufficient to initiate an impulse. This is the phenomenon of ***summation of impulses.***

Hyperpolarization and refractory period: Potassium channels close and short term conformational changes in the pore proteins of the sodium channels mean that these voltage-gated sodium channels are ***inactivated.*** As a result the neuronal membrane becomes ***refractory*** – unable to respond to a stimulus which would normally trigger an action potential.

Synapse: structure and function

Transmission of an action potential across a chemical synapse involves a uni-directional release of molecules of neurotransmitter from pre-synaptic to post-synaptic membranes

Drugs and poisons may interfere with synaptic transmission by

1. ***Mimicry of neurotransmitter,*** e.g. ***nicotine*** mimics both acetylcholine and noradrenaline
2. ***Reduced degradation,*** e.g. ***cocaine*** inhibits re-uptake of noradrenaline
3. ***Blocking receptors,*** e.g. ***chlorpromazine*** acts as an emotional depressant by blocking dopamine receptors
4. ***Reduced release of neurotransmitter,*** e.g. ***alcohol*** alters sleeping patterns by reducing release of serotonin.

1. **Increase in local Ca^{2+} concentration:** depolarization of membrane at synaptic button affects 'calcium channels' so that Ca^{2+} ions flow quickly into synaptic button from tissue fluid.

2. **Synaptic vesicles** containing molecules of neuro-transmitter move towards the presynaptic membrane.

Mitochondria are abundant in the synaptic button: release energy for refilling of synaptic vesicles and possibly for pumping of Ca^{2+} to re-establish Ca^{2+} concentration gradient across neurone membrane.

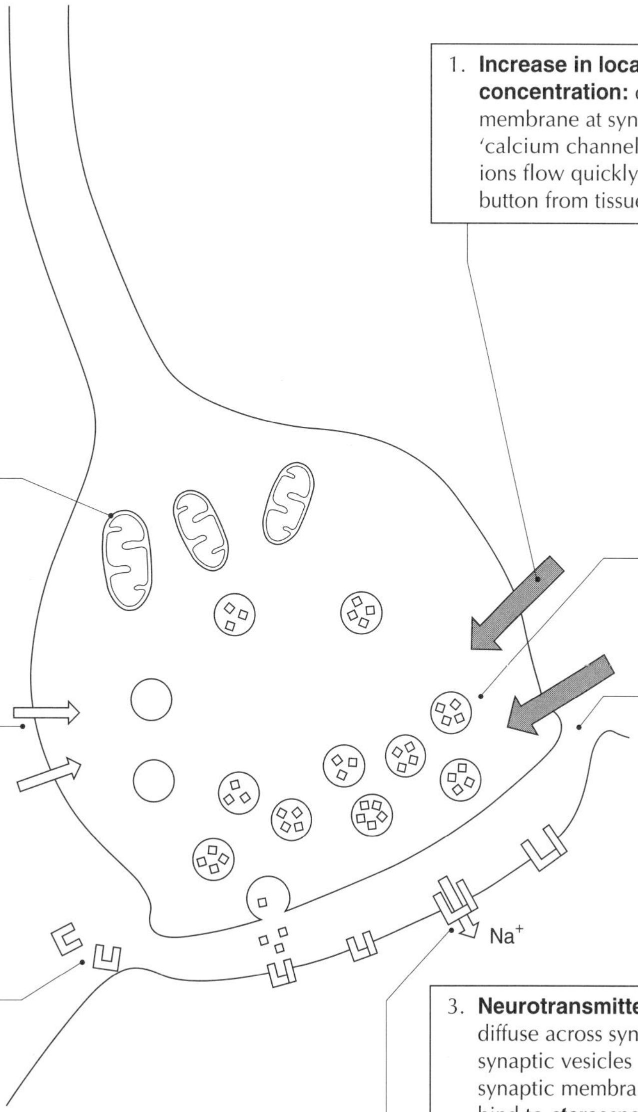

Synaptic cleft represents a barrier to the direct passage of the wave of depolarization from pre-synaptic to post-synaptic membranes.

6. **Reabsorption of neurotransmitter or products of degradation.** Molecules are resynthesized and reincorporated into synaptic vesicles. ***Catecholamines*** are often reabsorbed without degradation.

Na^+

5. **Enzymes degrade neurotransmitter.** These degradative enzymes, which are released from adjacent glial cells or are located on the post-synaptic membrane, remove neurotransmitter molecules so that their effect on the chemically gated ion channels is only short-lived. They include
   monoamine oxidase (degrades ***catecholamines)***
   acetylcholine esterase (degrades ***acetylcholine)***

4. **Chemically gated ion channels** on post-synaptic membrane – allow influx of Na^+ and efflux of K^+ → depolarization of post-synaptic membrane. Ion channels are 'opened' when triggered by binding of neurotransmitter.

3. **Neurotransmitter molecules** diffuse across synaptic gap when synaptic vesicles fuse with pre-synaptic membrane. Molecules bind to ***stereospecific receptors*** in the post-synaptic membrane. ***Catecholamines*** such as adrenaline are released from ***adrenergic nerve endings, acetylcholine*** from ***cholinergic nerve endings,*** and ***GABA*** (gamma-aminobutyric acid) and ***serotonin*** at synapses in the brain.

Excitatory post-synaptic potentials result if the neurotransmitter binding to the receptors on the post-synaptic membrane ***opens*** chemically gated ion channels, making ***depolarization more likely.***

Inhibitory post-synaptic potentials result if the neurotransmitter binding to the receptors on the post-synaptic membrane ***keeps*** chemically gated ion channels ***closed,*** promoting ***hyperpolarization*** and making ***depolarization less likely.***

Structure and function of the mammalian brain

Basal ganglia control gross muscle movements and regulate muscle tone. ***Limbic system*** controls involuntary elements of behaviour essential to survival, such as hissing and grimacing when threatened, or signs of pleasure during sexual activity.

Cerebral cortex has ***motor areas*** which control voluntary movement, ***sensory areas*** which interpret sensory information and ***association areas*** responsible for learning and emotion – connect and integrate sensory and motor regions.

CEREBRUM is the centre of intellect, memory, language and consciousness.

Corpus callosum is white matter and connects the right and left hemispheres of the brain.

Thalamus: main ***relay centre*** between cerebrum and spinal cord; preliminary sorting of incoming sensory impulses.

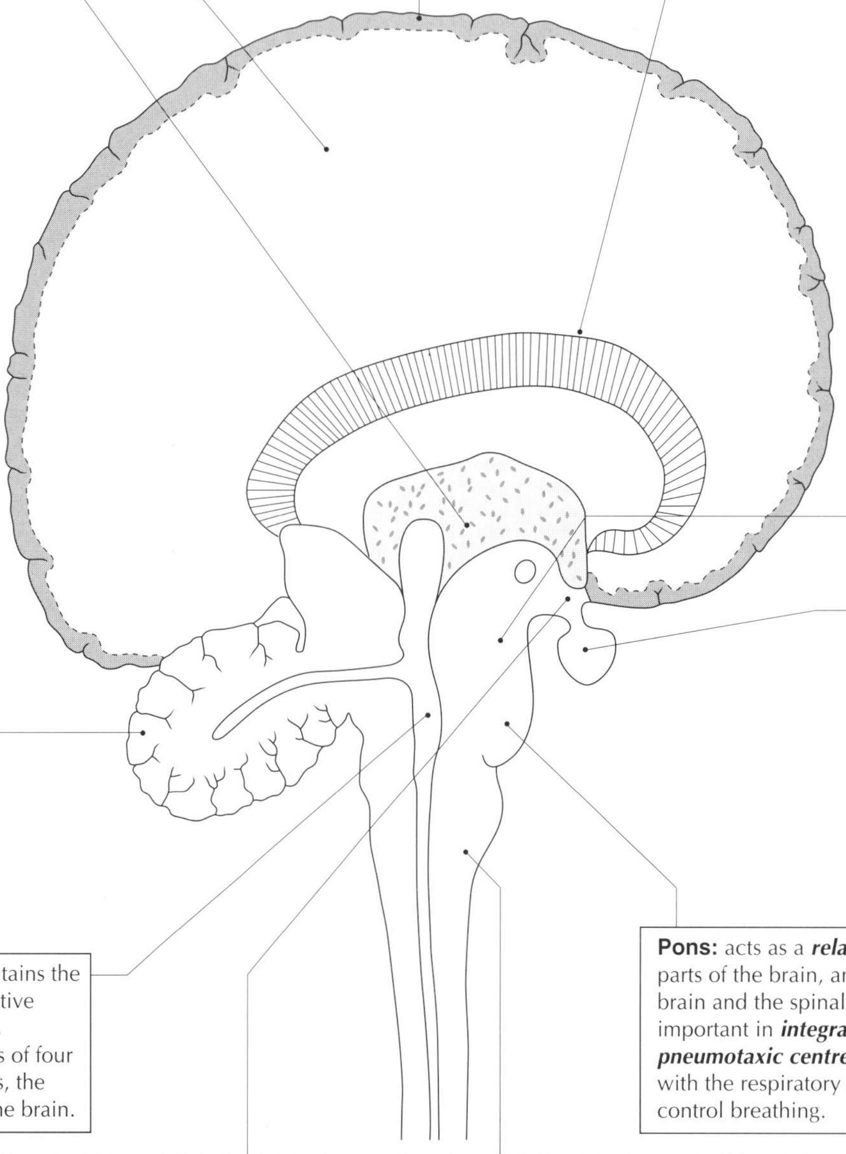

Midbrain is a ***relay centre*** but also contains structures – the ***colliculi*** – which control reflex movements of the head in response to visual and auditory stimuli. The ***substantia nigra*** secretes dopamine (degeneration → ***Parkinson's disease***) and the ***red nucleus*** integrates information from cerebrum and cerebellum concerning muscle tone and posture.

Cerebellum is responsible for smooth, co-ordinated movement via ***proprioception*** (the body's awareness of the position of one part relative to another). Maintains ***posture*** and ***muscle tone,*** and ***body equilibrium*** (using sensory input from inner ear). May regulate ***emotional development,*** modulating sensations of anger and pleasure.

Pituitary body (hypophysis) has a number of important endocrine functions, especially the secretion of ***trophins*** which regulate other endocrine organs.

Central canal: contains the protective and nutritive ***cerebrospinal fluid.*** Extends into a series of four CSF-filled chambers, the ***ventricles,*** within the brain.

Pons: acts as a ***relay centre*** between parts of the brain, and between the brain and the spinal cord– thus important in ***integration.*** Contains the ***pneumotaxic centre*** which, together with the respiratory centre, helps to control breathing.

Hypothalamus: contains centres for the control of body temperature (***thermoregulation***), hunger (***satiety***) and fluid balance (***osmoregulation***). Secretes ***releasing hormones*** which regulate the activity of the pituitary body (collectively called the ***hypothalamo-hypophysial system***) and two other hormones, ***oxytocin*** and ***antidiuretic hormone.*** It controls and integrates the ***autonomic nervous system,*** thus regulating heart rate, movement of food through the gut and contraction of the urinary bladder. It controls ***mind over body*** phenomena – it is the centre for many ***psychosomatic disorders*** and is associated with feelings of ***rage and aggression.*** It is one of the centres which control waking and sleeping patterns.

Medulla (medulla oblongata): acts as a ***relay centre*** for both sensory and motor impulses between other parts of the brain and the spinal cord. Has part of the ***reticular formation*** which functions in consciousness and arousal (the most common knockout blow hits the chin, twists and distorts the brain stem and overwhelms the ***reticular activating system*** by sending a rapid sequence of impulses → unconsciousness). Has ***vital reflex centres*** which regulate heart beat, breathing and blood vessel diameter, and ***nonvital reflex centres*** which co-ordinate swallowing, coughing, vomiting and sneezing. Also associated with much of the ***vestibular nuclear complex*** which is important in helping the body to maintain its sense of equilibrium.

Synovial joints have a space between the articulating bones and are freely movable.

THE MOVEMENTS POSSIBLE AT SYNOVIAL JOINTS ARE:

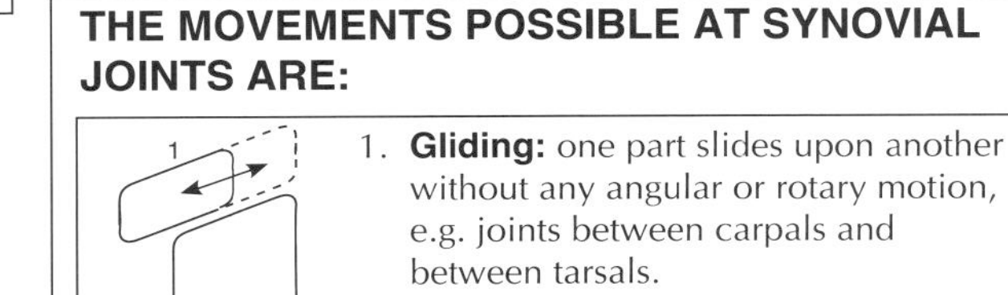

1. **Gliding:** one part slides upon another without any angular or rotary motion, e.g. joints between carpals and between tarsals.

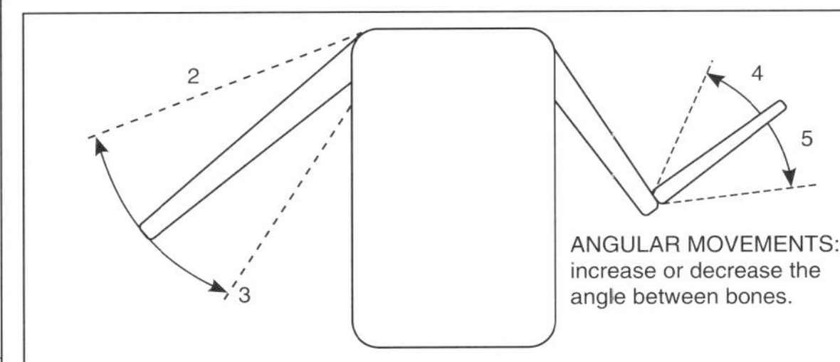

2. **Abduction:** moving the part away from the midline of the body.
3. **Adduction:** bringing the part towards the midline.
4. **Flexion:** decreasing the angle between two bones, includes bending the head forward (joint between the occipital bone and the atlas).
5. **Extension:** increasing the angle between two bones, includes returning the head to the anatomical position.

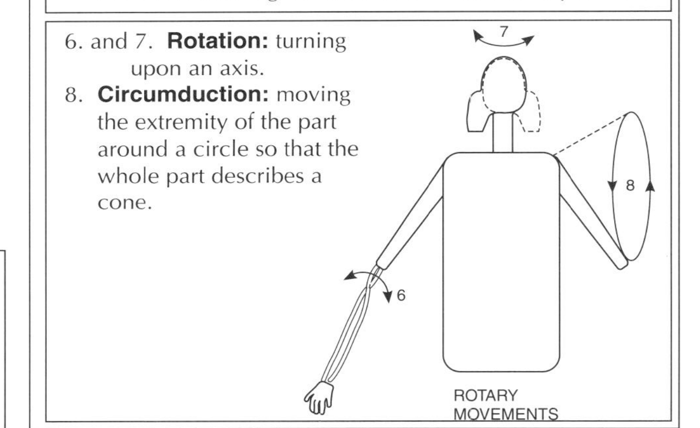

6. and 7. **Rotation:** turning upon an axis.

8. **Circumduction:** moving the extremity of the part around a circle so that the whole part describes a cone.

Fibrous capsule is formed of a connective tissue with an abundance of collagen fibres. It is attached to the periosteum of the articulating bones, and may be almost indistinguishable from extracapsular accessory ligament. The flexibility of the fibrous capsule permits movement at a joint and the great tensile strength of the collagen fibres resists dislocation.

Synovial membrane is internal to the fibrous capsule and is composed of loose connective tissue with an abundance of elastin fibres and variable amounts of adipose tissue. The membrane secretes ***synovial fluid*** which lubricates the joint and nourishes the articular cartilage.

Intracapsular ligament is located inside the articular capsule, but surrounded by folding of the synovial membrane. Examples are the ***cruciate ligaments*** of the knee joint.

Meniscal cartilage (mediscus, articular disc) is a pad of fibrocartilage which lies between the articular surfaces of the bones. By modifying the shapes of the articulating surfaces they allow two bones of different shapes to fit tightly together. Menisci also help to maintain the stability of the joint, and direct the flow of synovial fluid to areas of greatest friction.

Synovial fluid consists of hyaluronic acid and interstitial fluid formed from blood plasma. It is similar in appearance and consistency to egg white – in joints with little or no movement the fluid is viscous, but becomes less so as movement increases. The fluid also contains phagocytic cells which remove microbes and debris that results from wear and tear in the joint.

Articular cartilage of synovial joints is hyaline cartilage. It is able to reduce friction in the joint since it is coated by synovial fluid – as the load on the joint increases the spongy nature of the cartilage allows it to absorb water and so increase the proportion of the lubricating hyaluronic acid in the lubricating film. The cartilage may also function as a shock absorber although much of the 'impact loading' on a joint is probably absorbed by the thicker trabecular (honeycombed) bone which lies beneath it.

Extracapsular ligament is one of the possible ***accessory ligaments*** which offer additional stability to a joint. 'Extracapsular' means outside the articular capsule, although often the ligament is fused with the capsule. An example is the ***fibular collateral ligament*** of the knee joint.

Bursae are sac-like structures with walls of connective tissue lined by synovial membrane, and filled with a fluid similar in composition and function to synovial fluid. They are found between muscle and bone, tendon and bone and ligament and bone, and reduce the friction between one structure and another as movement takes place.

Movement of the forelimb illustrates the action of ***muscle groups.*** Skeletal muscles produce movement by exerting forces of contraction on tendons, which in turn pull on bones.

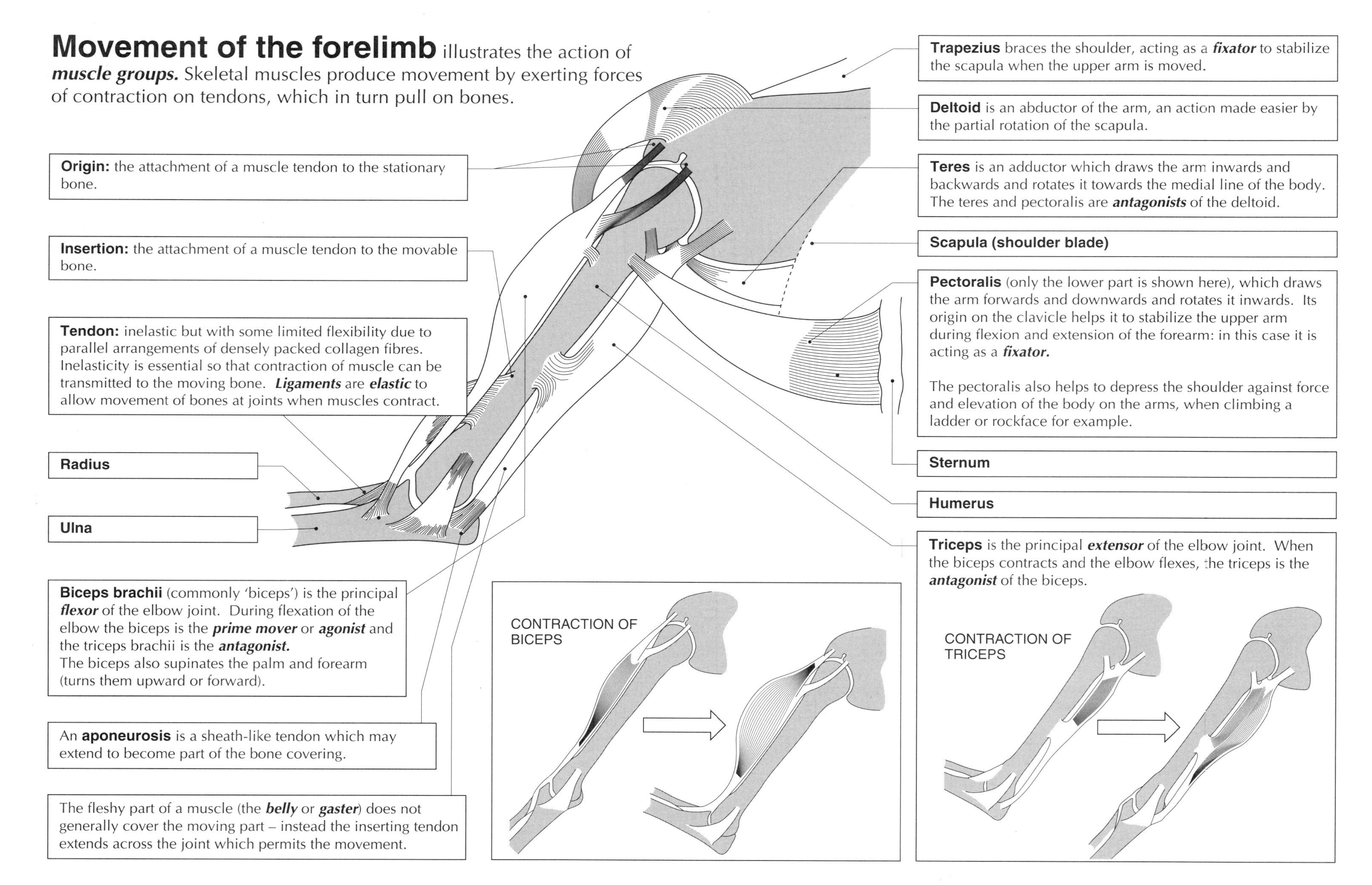

Origin: the attachment of a muscle tendon to the stationary bone.

Insertion: the attachment of a muscle tendon to the movable bone.

Tendon: inelastic but with some limited flexibility due to parallel arrangements of densely packed collagen fibres. Inelasticity is essential so that contraction of muscle can be transmitted to the moving bone. ***Ligaments*** are ***elastic*** to allow movement of bones at joints when muscles contract.

Radius

Ulna

Biceps brachii (commonly 'biceps') is the principal ***flexor*** of the elbow joint. During flexation of the elbow the biceps is the ***prime mover*** or ***agonist*** and the triceps brachii is the ***antagonist.***
The biceps also supinates the palm and forearm (turns them upward or forward).

An **aponeurosis** is a sheath-like tendon which may extend to become part of the bone covering.

The fleshy part of a muscle (the ***belly*** or ***gaster***) does not generally cover the moving part – instead the inserting tendon extends across the joint which permits the movement.

Trapezius braces the shoulder, acting as a ***fixator*** to stabilize the scapula when the upper arm is moved.

Deltoid is an abductor of the arm, an action made easier by the partial rotation of the scapula.

Teres is an adductor which draws the arm inwards and backwards and rotates it towards the medial line of the body. The teres and pectoralis are ***antagonists*** of the deltoid.

Scapula (shoulder blade)

Pectoralis (only the lower part is shown here), which draws the arm forwards and downwards and rotates it inwards. Its origin on the clavicle helps it to stabilize the upper arm during flexion and extension of the forearm: in this case it is acting as a ***fixator.***

The pectoralis also helps to depress the shoulder against force and elevation of the body on the arms, when climbing a ladder or rockface for example.

Sternum

Humerus

Triceps is the principal ***extensor*** of the elbow joint. When the biceps contracts and the elbow flexes, the triceps is the ***antagonist*** of the biceps.

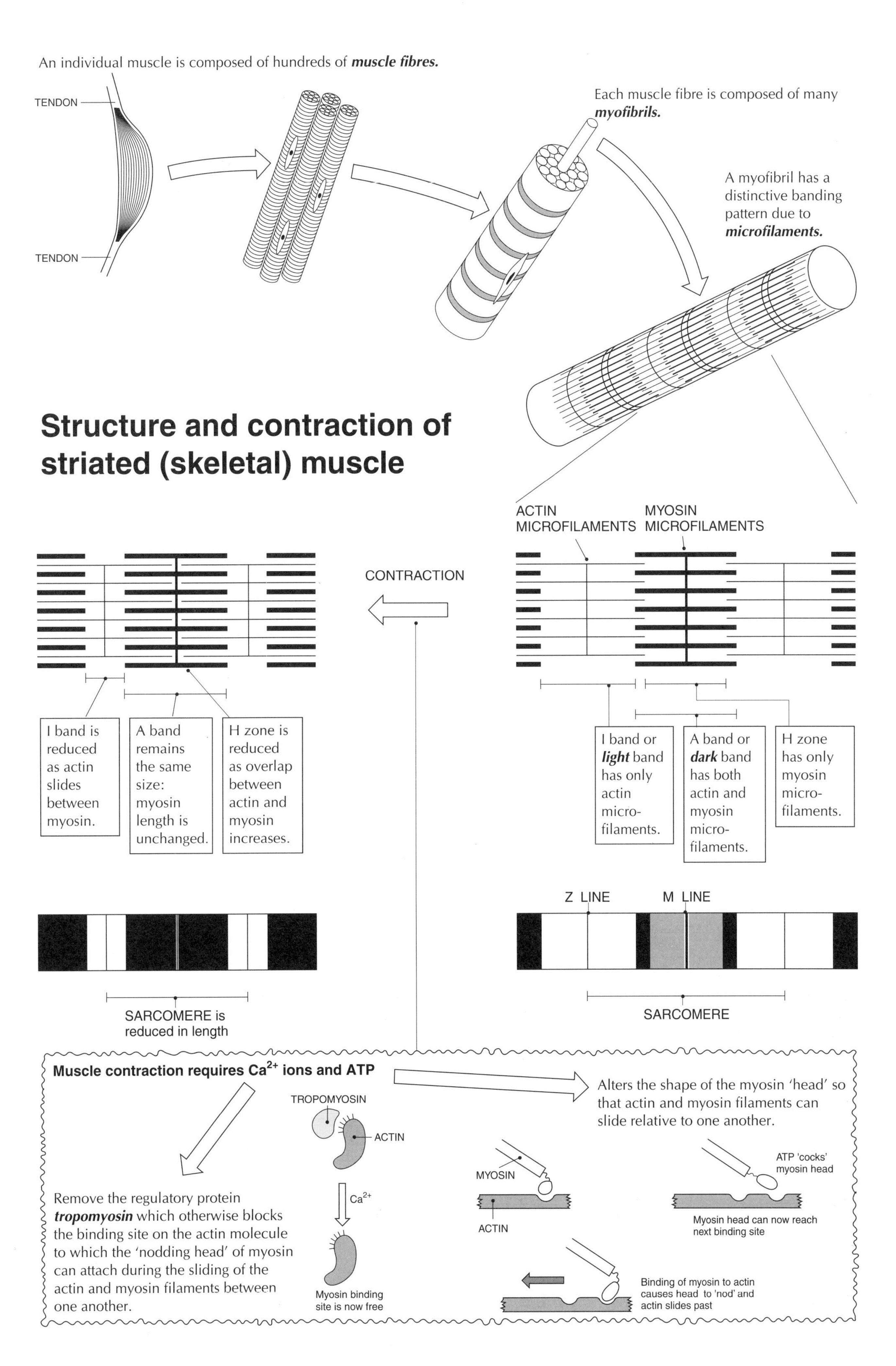
An individual muscle is composed of hundreds of muscle fibres.
TENDON
TENDON
Each muscle fibre is composed of many myofibrils.
A myofibril has a distinctive banding pattern due to microfilaments.
Structure and contraction of striated (skeletal) muscle
ACTIN MICROFILAMENTS
MYOSIN MICROFILAMENTS
CONTRACTION
I band is reduced as actin slides between myosin.
A band remains the same size: myosin length is unchanged.
H zone is reduced as overlap between actin and myosin increases.
I band or light band has only actin micro-filaments.
A band or dark band has both actin and myosin micro-filaments.
H zone has only myosin micro-filaments.
Z LINE
M LINE
SARCOMERE is reduced in length
SARCOMERE
Muscle contraction requires Ca2+ ions and ATP
TROPOMYOSIN
ACTIN
Ca2+
Myosin binding site is now free
Remove the regulatory protein tropomyosin which otherwise blocks the binding site on the actin molecule to which the 'nodding head' of myosin can attach during the sliding of the actin and myosin filaments between one another.
Alters the shape of the myosin 'head' so that actin and myosin filaments can slide relative to one another.
MYOSIN
ACTIN
ATP 'cocks' myosin head
Myosin head can now reach next binding site
Binding of myosin to actin causes head to 'nod' and actin slides past

Male reproductive system

The **vas deferens** stores sperm for up to several months and then propels them by peristalsis towards the urethra during ejaculation. Cutting and tying of the vas deferens constitutes ***vasectomy***, surgical sterilization of the male.

The **prostate gland** secretes an alkaline fluid which makes up about 20% of the volume of the semen – this alkalinity neutralizes the acidic secretions of the vagina, essential since sperm motility is considerably reduced at low pH. Cancer of the prostate is the most common tumour of the male reproductive system.

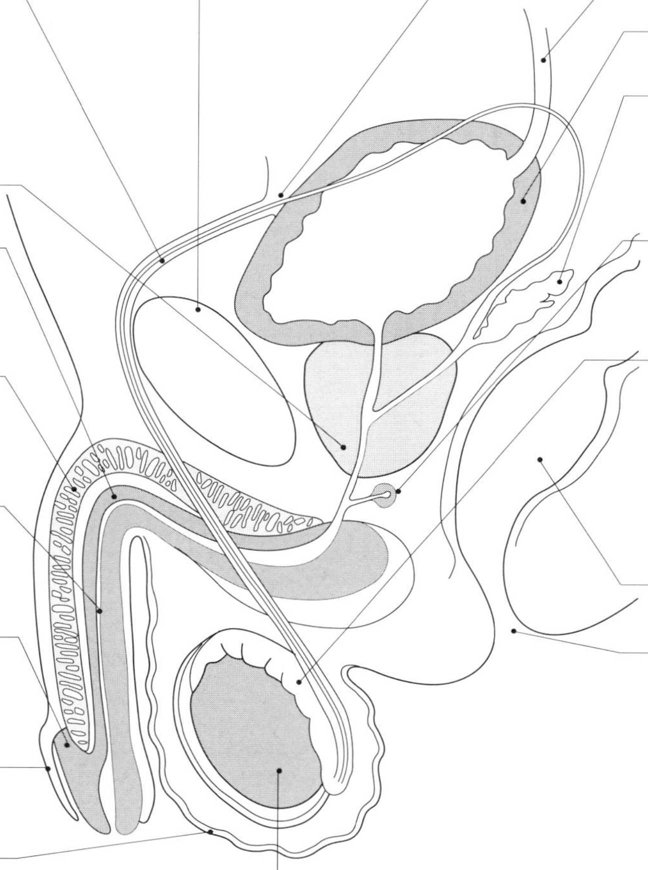

The **penis** is the intromittant organ, used to deliver sperm to the neck of the cervix, as close to the site of ovulation as possible. The penis is usually limp and flaccid but is erected by dilation of blood vessels allowing large quantities of blood to fill the sinuses in the ***corpora cavernosa and spongiosa.*** The ***glans penis*** has a high concentration of sensory cells and plays an important part in initiation of the erection reflex. The ***prepuce*** or ***foreskin*** covers the glans – it is sometimes removed by a surgical procedure called ***circumcision.***

Scrotum: the location of the scrotum and the contraction of specific muscle fibres regulate the temperature of the testes. This regulation is vital since sperm production and survival depend on a temperature about 3°C lower than body temperature, a temperature which is possible in the scrotum since this sac is outside the body cavities. The ***cremaster muscle*** elevates the testes when they are exposed to cold, moving them closer to the groin where they can absorb body heat. Exposure to a higher temperature reverses this process.

The **inguinal canal** is a service and supporting structure of the male reproductive system. It carries the vas deferens, testicular artery and vein and branches of the autonomic nervous system. The canal is a weak spot in the abdominal wall and is frequently the site of an inguinal hernia – a rupture of this region of the abdominal wall.

PUBIC BONE

URETER

BLADDER

CORPUS SPONGIOSUM

CORPUS CAVERNOSUM

URETHRA

GLANS PENIS

FORESKIN

RECTUM

ANUS

The **seminal vesicles** secrete an alkaline viscous fluid which makes up about 60% of the volume of the semen. The fluid is rich in fructose (the respiratory substrate for sperm motility) and is an important contributor to sperm viability.

The **Cowper's (bulbourethral) gland** produces an antacid fluid and also a mucus secretion which helps to lubricate the penis during intercourse.

The **epididymis** is the site of ***sperm maturation*** which begins in the seminiferous tubule, a process which takes between one and ten days. The epididymis also stores spermatozoa for up to four weeks, after which they are reabsorbed. The epididymis is lined with ciliated epithelium and has a smooth muscle layer – together these drive the sperm into the vas deferens.

The **testes** contain tightly coiled ***seminiferous tubules*** that produce sperm by a process called ***spermatogenesis.*** The interstitial tissue of the testes produces the hormone testosterone under the influence of LH from the anterior pituitary gland.

Semen (seminal fluid) is the fluid ejaculated from the urethra during sexual excitation. It is a mixture of sperm and the secretions of the prostate gland, Cowper's glands and the seminal vesicles. The sperm number about 250 000 000 per ejaculation, but constitute only about 1% of the total volume of between 2 and 5 cm^3 of the semen. The various secretions neutralize the acid environment of the male urethra and the female vagina and contain enzymes which are responsible for the final ***capacitation*** (activation) of the sperm. The semen also contains an antibiotic – ***seminalplasmin*** – which destroys some vaginal bacteria, and mucus to minimize friction between the penis and vagina during intercourse.

Human spermatozoon and oocyte

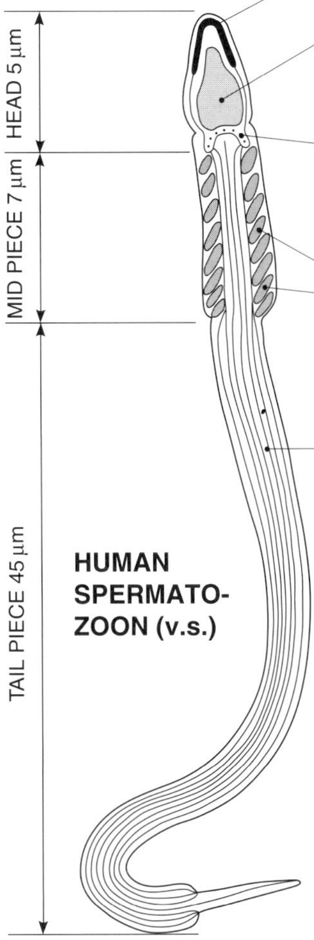

HEAD 5 μm

MID PIECE 7 μm

TAIL PIECE 45 μm

HUMAN SPERMATOZOON (v.s.)

Acrosome is effectively an enclosed lysosome. It develops from the Golgi complex and contains hydrolytic enzymes – a hyaluronidase and several proteinases – which aid in the penetration of the granular layer and plasma membrane of the oocyte immediately prior to fertilization.

Nucleus contains the haploid number of chromosomes derived by meiosis from the male germinal cells – thus this genetic complement will be either an X or a Y heterosome plus 22 autosomes. Since the head delivers no cytoplasm, the male contributes no extranuclear genes or organelles.

Centriole: one of a pair, which lie at right angles to one another. One of the centrioles produces microtubules which elongate and run the entire length of the rest of the spermatozoon, forming the axial filament of the flagellum.

Mitochondria are arranged in a spiral surrounding the flagellum. They complete the aerobic stages of respiration to release the ATP required for contraction of the filaments, leading to 'beating' of the flagellum and movement of the spermatozoon.

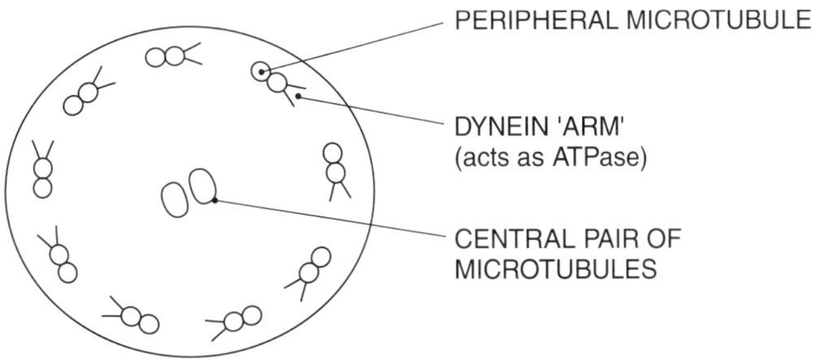

PERIPHERAL MICROTUBULE

DYNEIN 'ARM' (acts as ATPase)

CENTRAL PAIR OF MICROTUBULES

Flagellum has the 9 + 2 arrangement of microtubules typical of such structures. The principal role of the flagella is to allow sperm to move close to the oocyte and to orientate themselves correctly prior to digestion of the oocytemembranes. The sperm are moved close to the oocyte by muscular contractions of the walls of the uterus and the oviduct.

First polar body contains 23 chromosomes from the first meiotic division of the germ wall.

140 μm

Cumulus cells which once synthesized proteins and nucleic acids into the oocyte cytoplasm.

Zona pellucida will undergo structural changes at fertilization and form a barrier to the entry of more than one sperm.

23 chromosomes will complete second meiotic division on fertilization to provide ***female haploid nucleus*** and a second polar body (the ovum).

Cytoplasm may contribute extranuclear genes and organelles to the zygote.

Cortical granules contain enzymes which are released at fertilization and alter the structure of the zona pellucida, preventing further sperm penetration, which would upset the just restored diploid number.

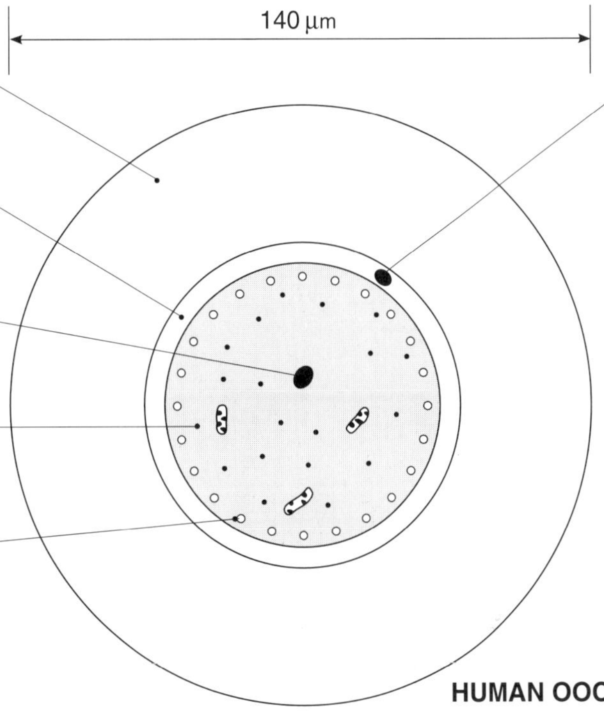

HUMAN OOCYTE (v.s.)

Female reproductive system

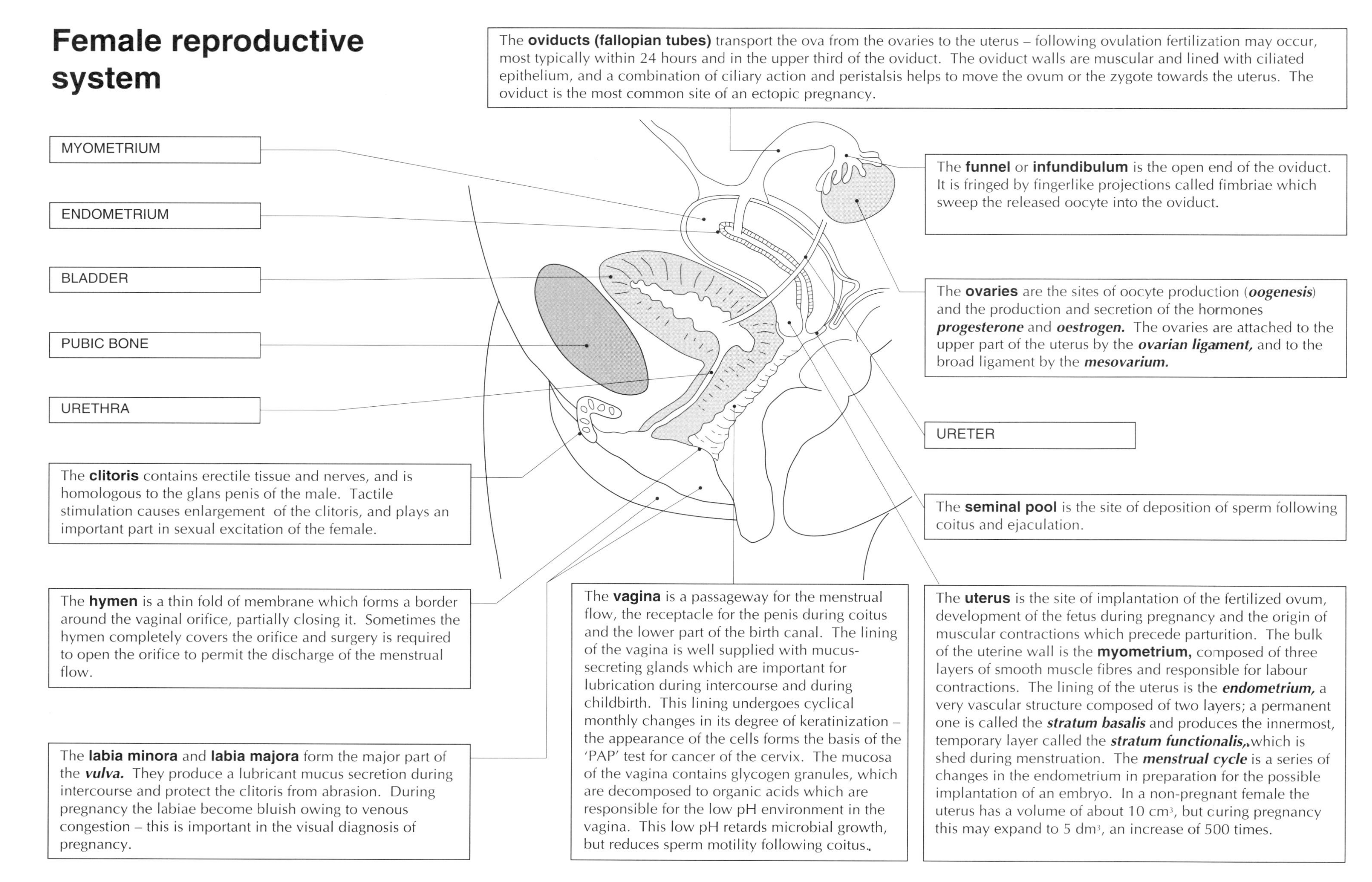

The **oviducts (fallopian tubes)** transport the ova from the ovaries to the uterus – following ovulation fertilization may occur, most typically within 24 hours and in the upper third of the oviduct. The oviduct walls are muscular and lined with ciliated epithelium, and a combination of ciliary action and peristalsis helps to move the ovum or the zygote towards the uterus. The oviduct is the most common site of an ectopic pregnancy.

The **funnel** or **infundibulum** is the open end of the oviduct. It is fringed by fingerlike projections called fimbriae which sweep the released oocyte into the oviduct.

The **ovaries** are the sites of oocyte production (***oogenesis***) and the production and secretion of the hormones ***progesterone*** and ***oestrogen.*** The ovaries are attached to the upper part of the uterus by the ***ovarian ligament,*** and to the broad ligament by the ***mesovarium.***

The **seminal pool** is the site of deposition of sperm following coitus and ejaculation.

The **clitoris** contains erectile tissue and nerves, and is homologous to the glans penis of the male. Tactile stimulation causes enlargement of the clitoris, and plays an important part in sexual excitation of the female.

The **hymen** is a thin fold of membrane which forms a border around the vaginal orifice, partially closing it. Sometimes the hymen completely covers the orifice and surgery is required to open the orifice to permit the discharge of the menstrual flow.

The **labia minora** and **labia majora** form the major part of the ***vulva.*** They produce a lubricant mucus secretion during intercourse and protect the clitoris from abrasion. During pregnancy the labiae become bluish owing to venous congestion – this is important in the visual diagnosis of pregnancy.

The **vagina** is a passageway for the menstrual flow, the receptacle for the penis during coitus and the lower part of the birth canal. The lining of the vagina is well supplied with mucus-secreting glands which are important for lubrication during intercourse and during childbirth. This lining undergoes cyclical monthly changes in its degree of keratinization – the appearance of the cells forms the basis of the 'PAP' test for cancer of the cervix. The mucosa of the vagina contains glycogen granules, which are decomposed to organic acids which are responsible for the low pH environment in the vagina. This low pH retards microbial growth, but reduces sperm motility following coitus.

The **uterus** is the site of implantation of the fertilized ovum, development of the fetus during pregnancy and the origin of muscular contractions which precede parturition. The bulk of the uterine wall is the **myometrium,** composed of three layers of smooth muscle fibres and responsible for labour contractions. The lining of the uterus is the ***endometrium,*** a very vascular structure composed of two layers; a permanent one is called the ***stratum basalis*** and produces the innermost, temporary layer called the ***stratum functionalis,*** which is shed during menstruation. The ***menstrual cycle*** is a series of changes in the endometrium in preparation for the possible implantation of an embryo. In a non-pregnant female the uterus has a volume of about 10 cm^3, but during pregnancy this may expand to 5 dm^3, an increase of 500 times.

Events of the menstrual cycle

Follicle-stimulating hormone (FSH) initiates the development of several primary follicles (each containing a primary oocyte): one follicle continues to develop but the others degenerate by the process of follicular atresia. FSH also increases the activity of the enzymes responsible for formation of oestrogen.

Oestrogen is produced by enzyme modification (in the stroma) of testosterone produced by the thecae of the developing follicle. Oestrogen has several effects:

1. It stimulates further growth of the follicle.
2. It promotes repair of the endometrium.
3. It acts as a feedback inhibitor of the secretion of FSH from the anterior pituitary gland.
4. From about day 11 it has a positive feedback action on the secretion of both LH and FSH.

Development of the follicle within the ovary is initiated by FSH but continued by LH. The Graafian follicle (Ⓐ) is mature by day 10–11 and ovulation occurs at day 14 (Ⓑ) following a surge of LH. The remains of the follicle become the corpus luteum (Ⓒ), which secretes steroid hormones. These steroid hormones inhibit LH secretion so that the corpus luteum degenerates and becomes the corpus albicans (Ⓓ).

The **endometrium** begins to thicken and become more vascular under the influence of the ovarian hormone oestrogen. Because of this thickening, to 4–6 mm, the time between menstruation and ovulation is sometimes called the ***proliferative phase.***

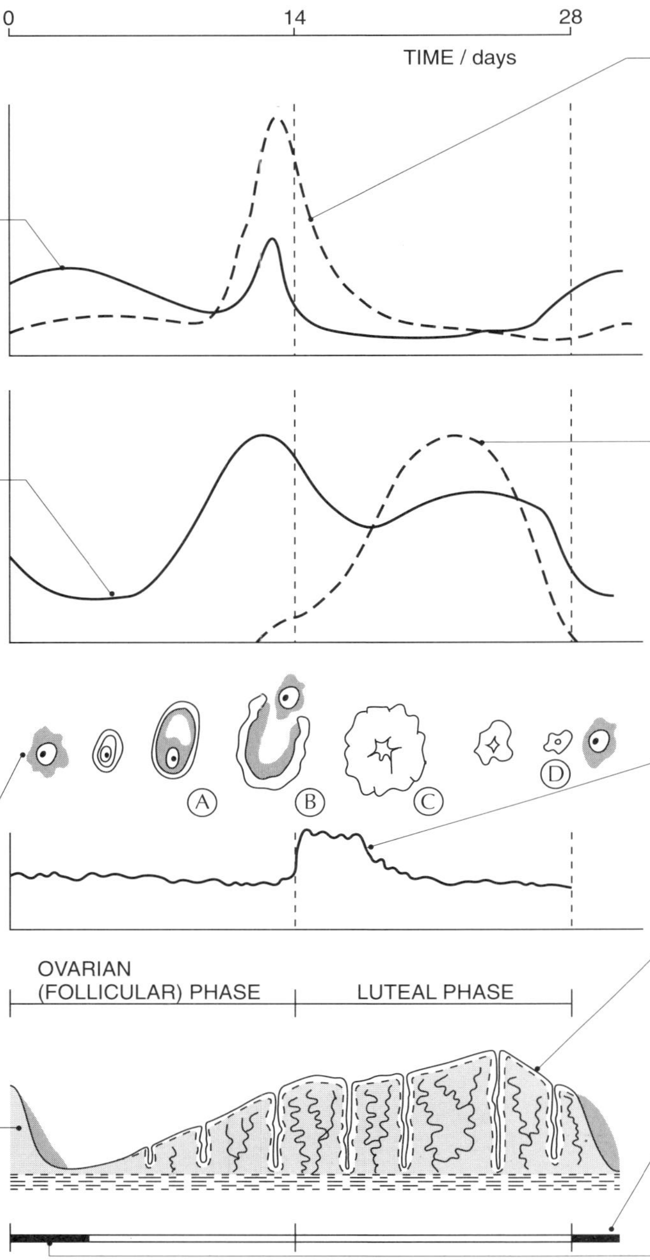

Luteinizing hormone (LH) triggers the secretion of testosterone by the thecae of the follicle, and when its concentration 'surges' it causes release of enzymes which rupture the wall of the ovary, allowing the secondary oocyte to be released at ovulation. After ovulation LH promotes development of the corpus luteum from the remains of the Graafian follicle.

Progesterone is secreted by the corpus luteum. It has several effects:

1. It prepares the endometrium for implantation of a fertilized egg by increasing vascularization, thickening and the storage of glycogen.
2. It begins to promote growth of the mammary glands.
3. It acts as a feedback inhibitor of FSH secretion, thus arresting development of any further follicles.

Body temperature rises by about 1°C at the time of ovulation. This 'heat' is used to determine the 'safe period' for the rhythm method of contraception.

During the post-ovulatory or luteal phase the **endometrium** becomes thicker with more tortuously coiled glands and greater vascularization of the surface layer, and retains more tissue fluid.

Menstruation is initiated by falling concentrations of oestrogen and progesterone as the corpus luteum degenerates.

At **menstruation** the ***stratum functionalis*** of the endometrium is shed, leaving the ***stratum basilis*** to begin proliferation of a new *functionalis.*

Functions of the placenta

Barrier: the placenta limits the transfer of solutes and blood components from maternal to fetal circulation. Cells of the maternal immune system do not cross – this minimizes the possibility of immune rejection (although antibodies may cross which may cause haemolysis of fetal blood cells if ***Rhesus*** antibodies are present in the maternal circulation). The placenta is ***not*** a barrier to heavy metals such as lead, to nicotine, to HIV and other viruses, to heroin and to toxins such as DDT. Thus the ***Rubella*** (German measles) virus may cross and cause severe damage to eyes, ears, heart and brain of the fetus, the sedative Thalidomide caused severely abnormal limb development, and some children are born already addicted to heroin or HIV positive.

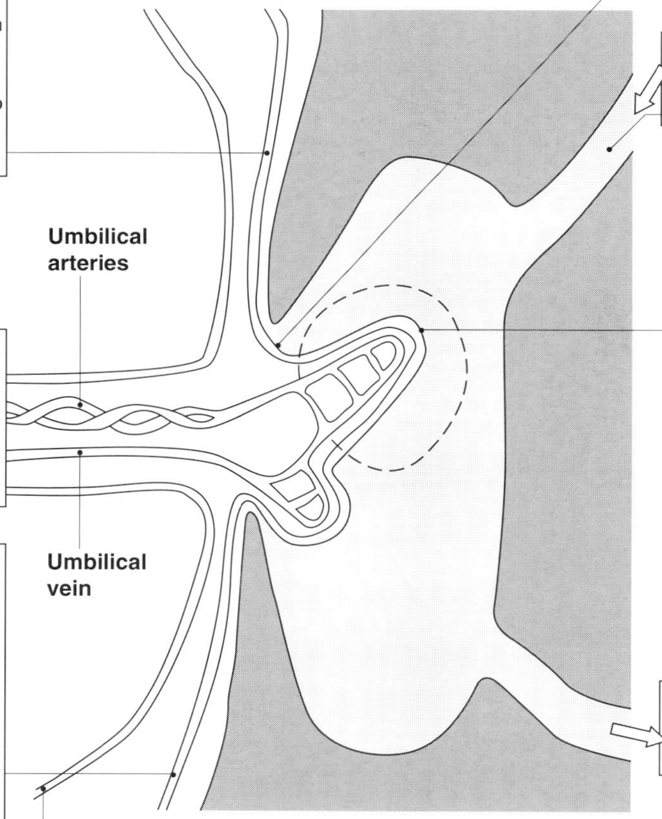

Immune protection: protective molecules (possibly including HCG) cover the surface of the early placenta and 'camouflage' the embryo which, with its complement of paternal genes, might be identified by the maternal immune system and rejected as tissue of 'non-self' origin.

Branch of uterine artery delivers maternal blood to the lacunae. Blood transports oxygen, soluble nutrients, hormones and antibodies, but also drugs and viruses.

Umbilical cord provides the connection between the abdomen of the foetus and the placenta. It is composed of the two spiralling umbilical arteries (carrying deoxygenated blood from the foetal circulation) and the umbilical vein (returning oxygenated, nutrient-enriched blood to the foetus) embedded in Wharton's jelly.

Chorionic villi are the sites of exchange of many solutes between maternal and fetal circulations. Oxygen transfer to the fetus is aided by fetal haemoglobin with its high oxygen affinity, and soluble nutrients such as glucose and amino acids are selectively transported by membrane-bound carrier proteins. Carbon dioxide and urea diffuse from fetus to mother along diffusion gradients maintained over larger areas of the placenta by countercurrent flow of fetal and maternal blood. In the later stages of pregnancy antibodies pass from mother to fetus – these confer immunity in the young infant, particularly to some gastro-intestinal infections. **There is no direct contact between maternal and foetal blood.**

Endocrine function: cells of the **chorion** secrete a number of hormones:

1. **HCG (human chorionic gonadotrophin)** maintains the corpus luteum so that this body may continue to secrete the progesterone necessary to continue the development of the endometrium. HCG is principally effective during the first 3 months of pregnancy and the overflow of this hormone into the urine is used in diagnosis of the pregnant condition.
2. **Oestrogen** and **progesterone** are secreted as the production of these hormones from the degenerating corpus luteum diminishes.
3. **Human placental lactogen** promotes milk production in the mammary glands as birth approaches.
4. **Prostaglandins** are released under the influence of fetal adrenal steroid hormones. These prostaglandins are powerful stimulators of contractions of the smooth muscle of the uterus, the contractions which constitute labour and eventually expel the fetus from the uterus.

Branch of uterine vein removes blood from lacunae. Blood contains increased concentrations of carbon dioxide, urea and placental hormones.

Amnion is an extra-embryonic membrane which surrounds the umbilical cord and encloses the **amniotic fluid** to form the **amniotic sac**. This fluid supports the embryo and provides samples of embryonic cells via the technique of **amniocentesis**. Rupture of the amnion, and release of the 'waters' is often the first sign that parturition is imminent.

After expulsion of the fetus, further contractions of the uterus cause detachment of the placenta (spontaneous constriction of uterine artery and vein limit blood loss) – the placenta, once delivered, is referred to as the ***afterbirth.*** The placenta may be used as a source of hormones (e.g. in 'fertility pills'), as tissue for burn grafts and to supply veins for blood vessel grafts.

DNA replication and chromosomes

Chromosomes contain genes (sections of DNA which code for particular protein products responsible for measurable characteristics).

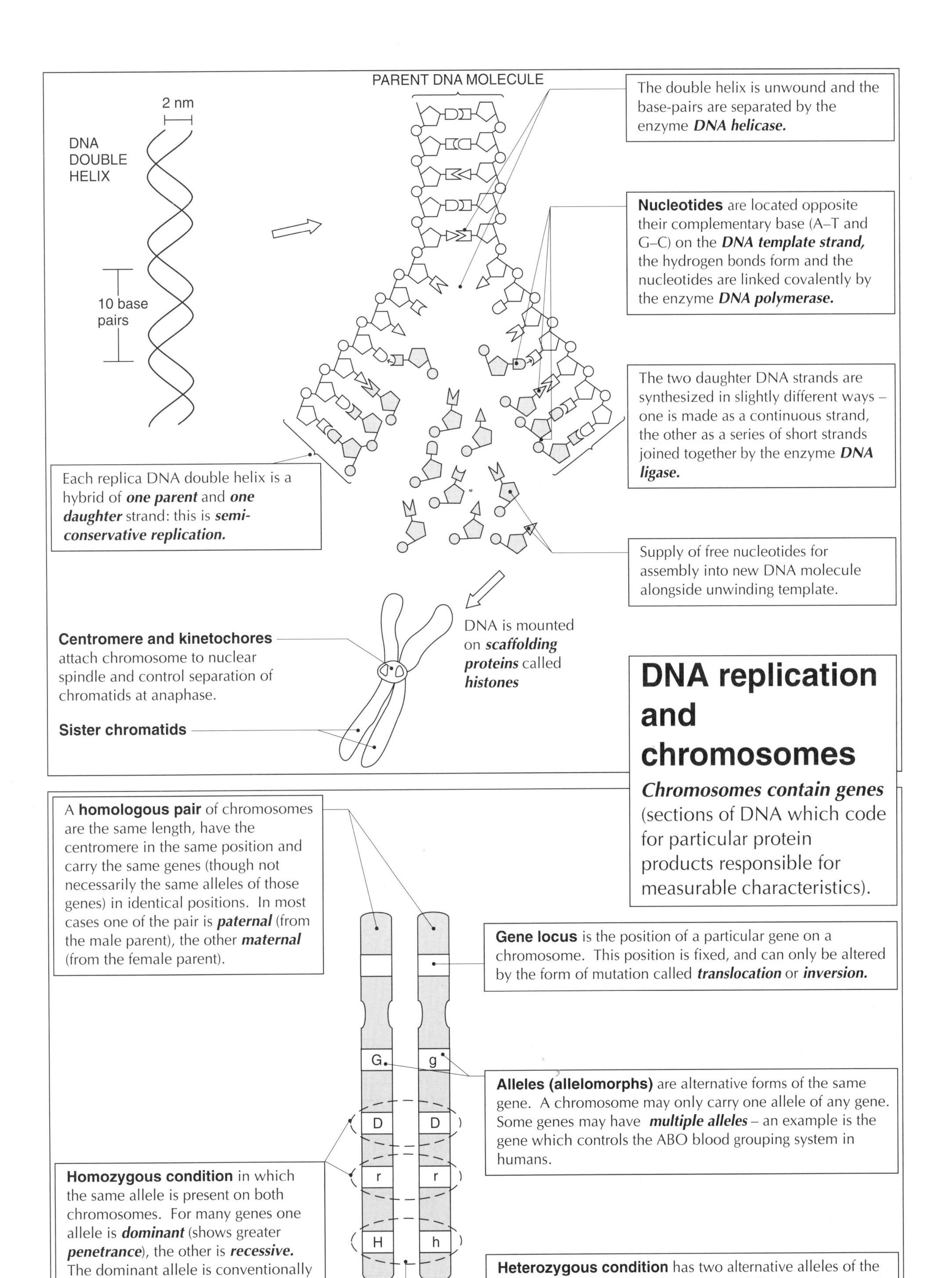

The double helix is unwound and the base-pairs are separated by the enzyme ***DNA helicase.***

Nucleotides are located opposite their complementary base (A–T and G–C) on the ***DNA template strand,*** the hydrogen bonds form and the nucleotides are linked covalently by the enzyme ***DNA polymerase.***

The two daughter DNA strands are synthesized in slightly different ways – one is made as a continuous strand, the other as a series of short strands joined together by the enzyme ***DNA ligase.***

Each replica DNA double helix is a hybrid of ***one parent*** and ***one daughter*** strand: this is ***semi-conservative replication.***

Supply of free nucleotides for assembly into new DNA molecule alongside unwinding template.

Centromere and kinetochores attach chromosome to nuclear spindle and control separation of chromatids at anaphase.

Sister chromatids

DNA is mounted on ***scaffolding proteins*** called ***histones***

A **homologous pair** of chromosomes are the same length, have the centromere in the same position and carry the same genes (though not necessarily the same alleles of those genes) in identical positions. In most cases one of the pair is ***paternal*** (from the male parent), the other ***maternal*** (from the female parent).

Gene locus is the position of a particular gene on a chromosome. This position is fixed, and can only be altered by the form of mutation called ***translocation*** or ***inversion.***

Alleles (allelomorphs) are alternative forms of the same gene. A chromosome may only carry one allele of any gene. Some genes may have ***multiple alleles*** – an example is the gene which controls the ABO blood grouping system in humans.

Homozygous condition in which the same allele is present on both chromosomes. For many genes one allele is ***dominant*** (shows greater ***penetrance***), the other is ***recessive.*** The dominant allele is conventionally written with an upper case symbol (here D) and the recessive allele with a lower case symbol (here r). In this example DD represents the ***homozygous dominant*** condition, rr the ***homozygous recessive*** condition.

Heterozygous condition has two alternative alleles of the same gene on the two chromosomes of the homologous pair. When one of the alleles is dominant over the other, only the dominant form will be expressed in the ***phenotype*** (the sum of the measurable characteristics of the individual, i.e. the two ***genotypes HH*** and ***Hh*** would have identical phenotypes.

Genes control cell characteristics

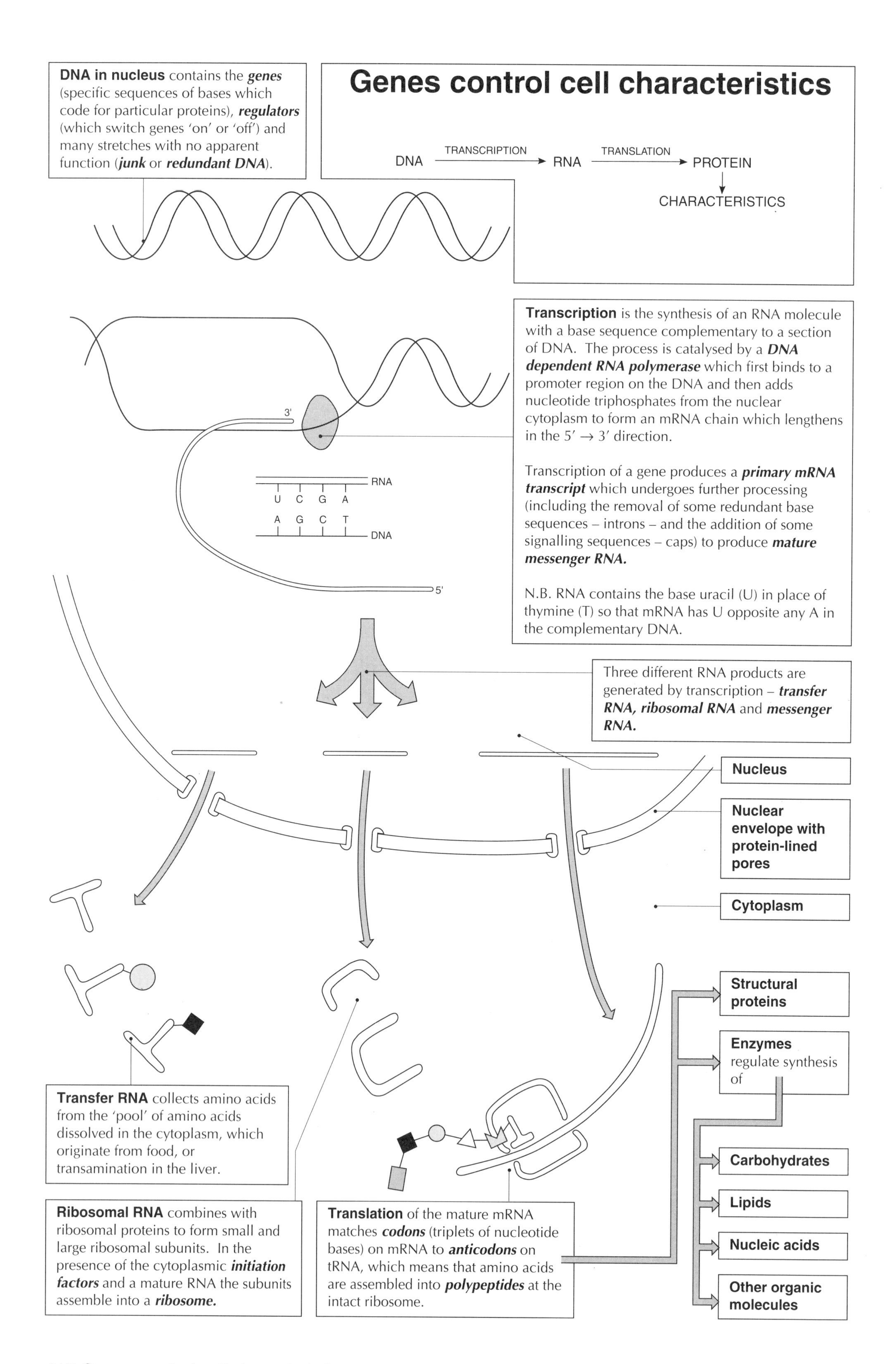

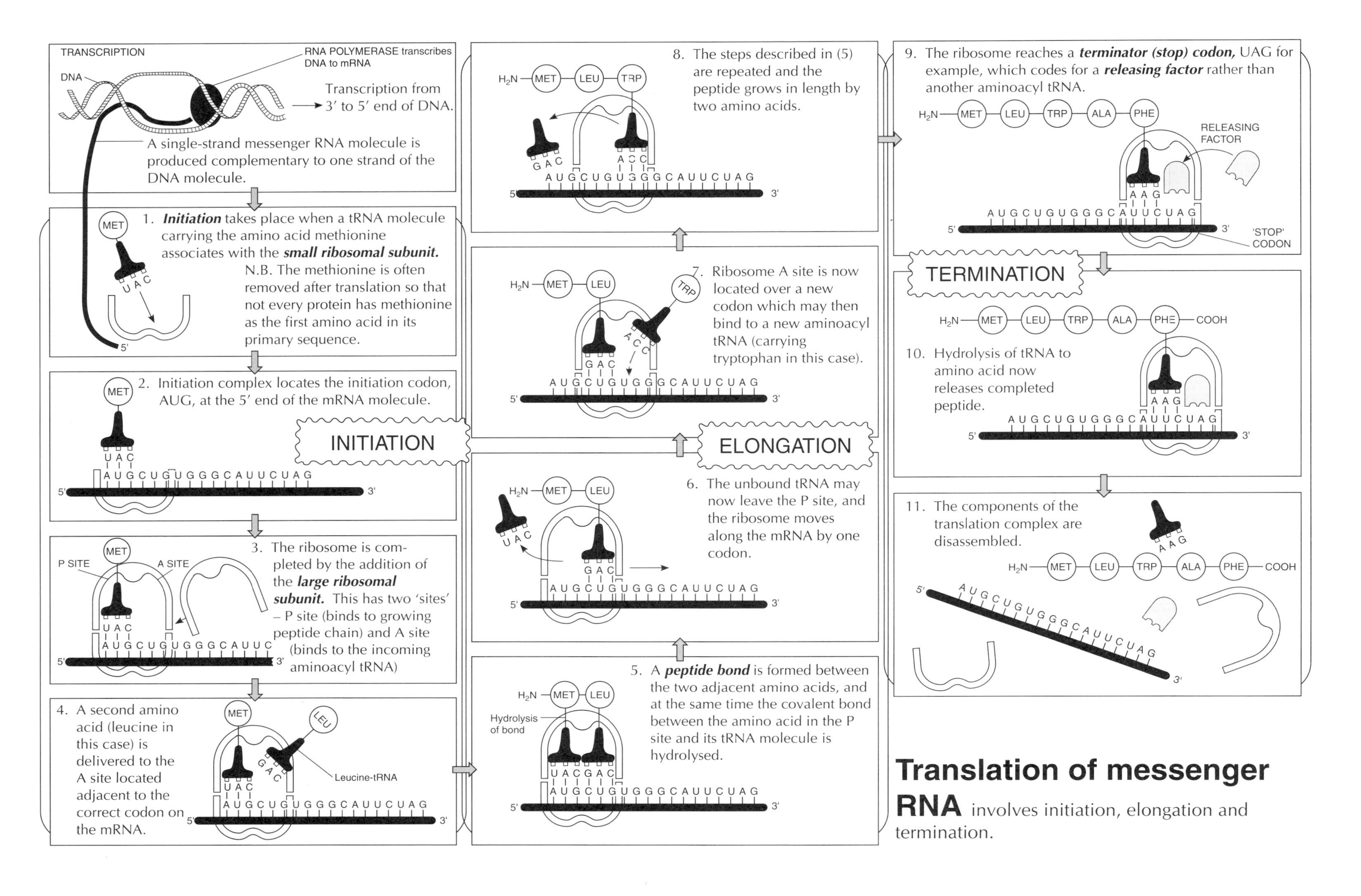

Translation of messenger RNA

involves initiation, elongation and termination.

Mitosis and growth

The significance of mitosis is that it involves duplication of the genetic material and its equal distribution to each of two 'daughter' cells: ***variation is minimal.***

Importance of mitosis

1. It is the process which provides the cells required for the ***growth*** of zygote into a functioning multicellular organism – this requires an increase in number of cells from one to 6 x 10^{13} in humans.
2. It supplies the cells to ***repair*** worn-out or damaged tissues. In the human the replacement of skin, gut and lung linings and blood cells requires about 1 x 10^{11} cells per day.
3. It maintains the chromosome number. Daughter cells have identical sets of chromosomes and so function harmoniously as part of the tissue, organ or organism.
4. ***Asexual reproduction*** provides offspring which are genetically identical to the parent – ideal when rapidly establishing a population. Mitosis provides the cells which make up the fragments of the parent body dispersed during this form of reproduction.

The cell cycle typically lasts from 8 to 24 h in humans – the nuclear division (mitosis) occupies about 10% of this time.

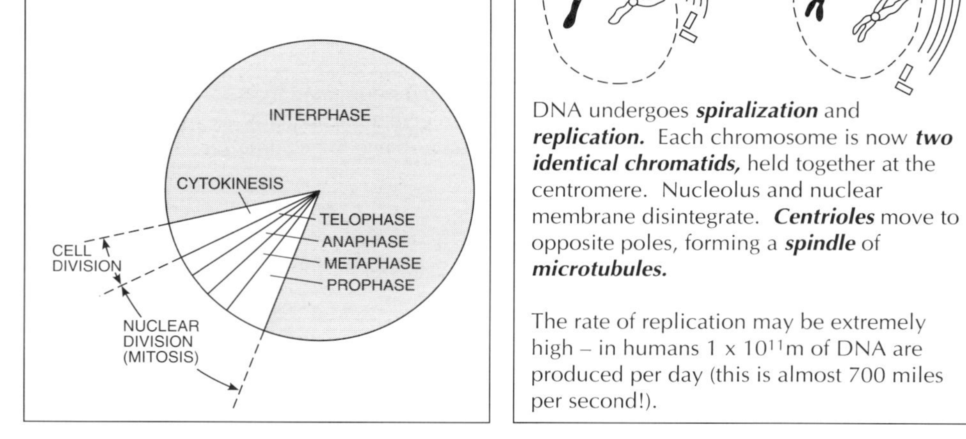

Interphase

DNA only visible as indistinct mass of ***chromatin.*** Nucleolus and nuclear membrane still intact. Centrioles lie close to one another.

4*n*

DNA CONTENT OF CELL

2*n*

As **interphase** moves to **prophase**

DNA undergoes ***spiralization*** and ***replication.*** Each chromosome is now ***two identical chromatids,*** held together at the centromere. Nucleolus and nuclear membrane disintegrate. ***Centrioles*** move to opposite poles, forming a ***spindle*** of ***microtubules.***

The rate of replication may be extremely high – in humans 1 x 10^{11}m of DNA are produced per day (this is almost 700 miles per second!).

At **metaphase**

Chromosomes now attached to spindle at ***kinetochore*** on the centromere. The chromosomes are arranged in such a way that one chromatid from each pair lies on each side of the ***equator.***

Anaphase precedes **telophase**

Centromere divides and spindle fibres contract to pull ***chromatids*** to opposite poles. The early separation of the chromatids constitutes ***anaphase,*** and the separation is complete (so that the chromatids are now ***chromosomes***) when the spindle disintegrates and the nuclear membrane reforms at ***telophase.***

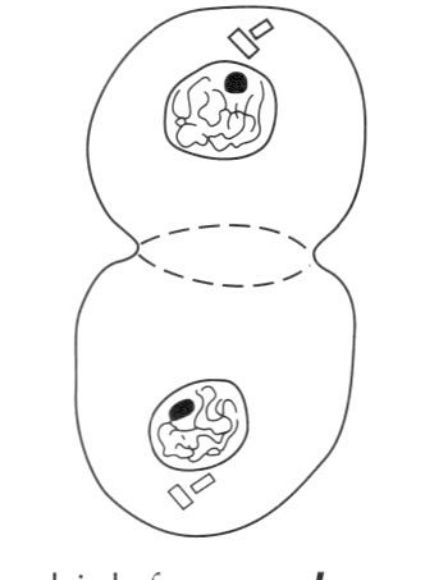

During ***cytokinesis*** the tetraploid (4*n*) cell is 'pinched' into two 'daughter cells'. Each product has a ***DNA content equal to the other and to the parent cell.*** In animal cells the separation is brought about by two contractile proteins which form a ***cleavage furrow;*** in plant cells a ***cell plate*** is laid down and covered with cellulose to form a separating ***cell wall.***

Meiosis and variation

Meiosis separates chromosomes, halving the diploid number, and introduces variation to the haploid products.

During **prophase I** each replicated chromosome (comprising two chromatids) pairs with its ***homologous partner,*** i.e. the diploid number of chromosomes produces the haploid number of homologous pairs.

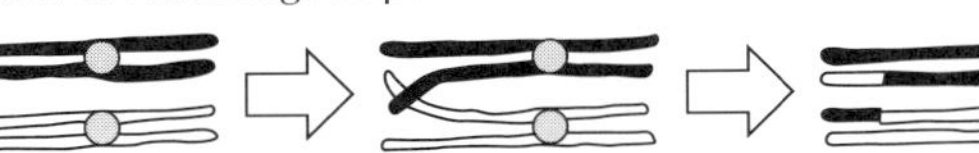

Crossing over occurs when all four chromatids are at ***synapsis*** (exactly aligned) – non-sister chromatids may cross over, break and reassemble so that ***parental*** gene combinations are replaced by ***recombinants.*** This is a major source of ***genetic variation.***

At **anaphase I** and **telophase I** there is separation of ***whole chromosomes*** (i.e. of ***pairs*** of chromatids).

The products of meiosis I now contain the ***haploid*** (n) number of chromosomes, although each chromosome comprises two chromatids.

During the **second meiotic division** (metaphase II, anaphase II and telophase II) there is a modified mitosis which ***separates the two sister chromatids of each chromosome.***

AND

DNA CONTENT OF CELL

$4n$

$2n$

n

Following prophase I, **independent assortment** can align the chromosomes in different ways on the ***metaphase*** plate.

OR

The number of possible combinations of chromosomes is great i.e. 2^n, where n is the number of homologous pairs. This is a second major source of ***genetic variation*** resulting from meiotic division.

At the end of telophase II **cytokinesis** produces daughter nuclei which have ***half the number of chromosomes of the parent cell.***

i.e. DIPLOID ($2n$) ⟶ HAPLOID (n) GAMETES

In addition, the unpaired chromosomes in the gametes may contain ***new gene combinations*** as a result of ***crossing over*** and ***independent assortment.***

Further genetic variation results from the ***random combination of gametes at fertilization,*** i.e. any male gamete may fuse with any female gamete.

Importance of meiosis

1. It must occur in sexually reproducing organisms or the chromosome number would be doubled at fertilization.

♂ PARENT ($2n$) —MEIOSIS→ ♂ GAMETE (n)

♀ PARENT ($2n$) —MEIOSIS→ ♀ GAMETE (n)

→ ZYGOTE ($2n$) → NEW INDIVIDUAL ($2n$)

2. Crossing over, independent assortment and random fertilization promote ***genetic variation.*** This provides new material for natural selection to work on during evolution.

Gene mutation and sickle cell anaemia

Sickle cell anaemia is the result of a ***single gene (point) mutation*** and the resulting ***error in protein synthesis.***

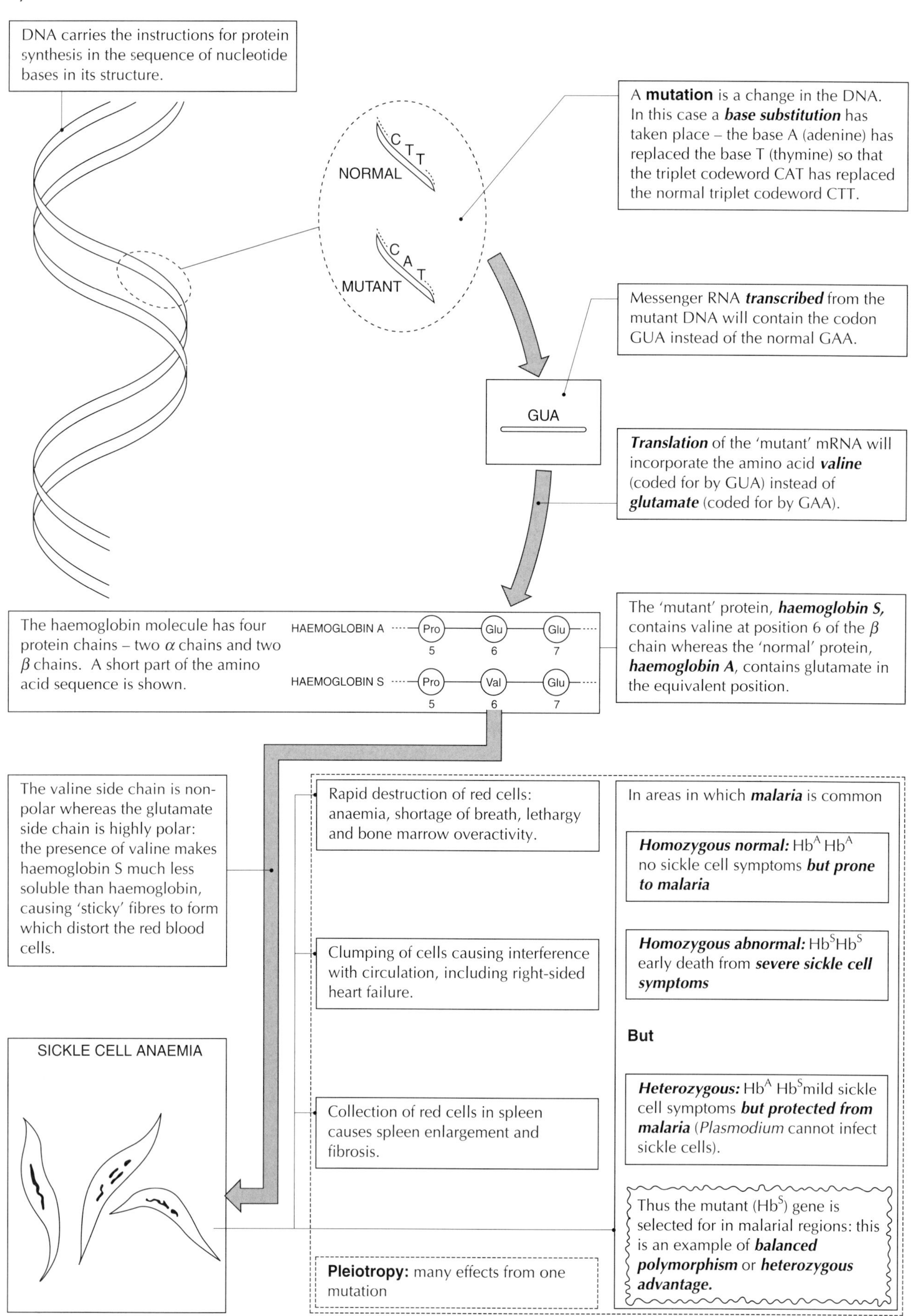

Chromosome mutation and Down's Syndrome

Down's syndrome (trisomy-21) is a chromosome mutation caused by ***non-disjunction.***

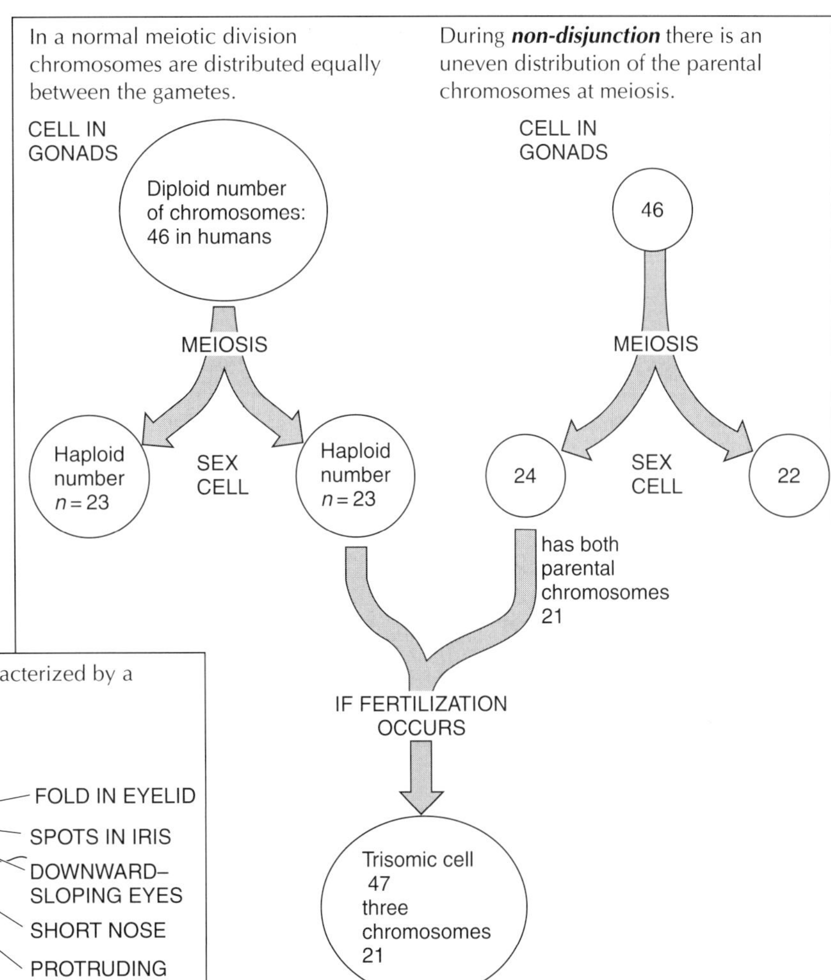

Down's syndrome (trisomy-21) is characterized by a number of distinctive physical features.

BROAD FOREHEAD
FOLD IN EYELID
SPOTS IN IRIS
DOWNWARD–SLOPING EYES
SHORT NOSE
PROTRUDING TONGUE
SHORT NECK

In addition there are congenital heart defects (30% of sufferers die before the age of 10) and mental retardation.

A **karyotype** is obtained by cutting out and rearranging photographic images of chromosomes stained during mitotic metaphase.

KARYOTYPE OF DOWN'S SYNDROME FEMALE

1 2 3 4 5
6 7 8 9 10 11 12
13 14 15 16 17 18 19 20
21 22 X Y

NOTE THE EXTRA CHROMOSOME 21

INCIDENCE OF DOWN'S SYNDROME
0 20 30 40 50
MOTHER'S AGE / years

The non-disjunction is most usually the result of the failure of chromosomes to separate at anaphase I – the probability of this happening increases with the length of time the cell remains in prophase I. In the human female all meioses are initiated before puberty so there is an age-related incidence of Down's syndrome – the longer the oocyte takes to complete development the less accurate are the chromosome separations which follow.

Paternal non-disjunction accounts for only about 15% of cases of Down's syndrome.

Two other significant examples of non-disjunction

Klinefelter's syndrome (XXY) caused by an ***extra X chromosome*** and resulting in a ***sterile male*** with ***some breast development.***

Turner's syndrome (XO) caused by a ***deleted X chromosome*** and resulting in a ***female*** with ***underdeveloped sexual characteristics.***

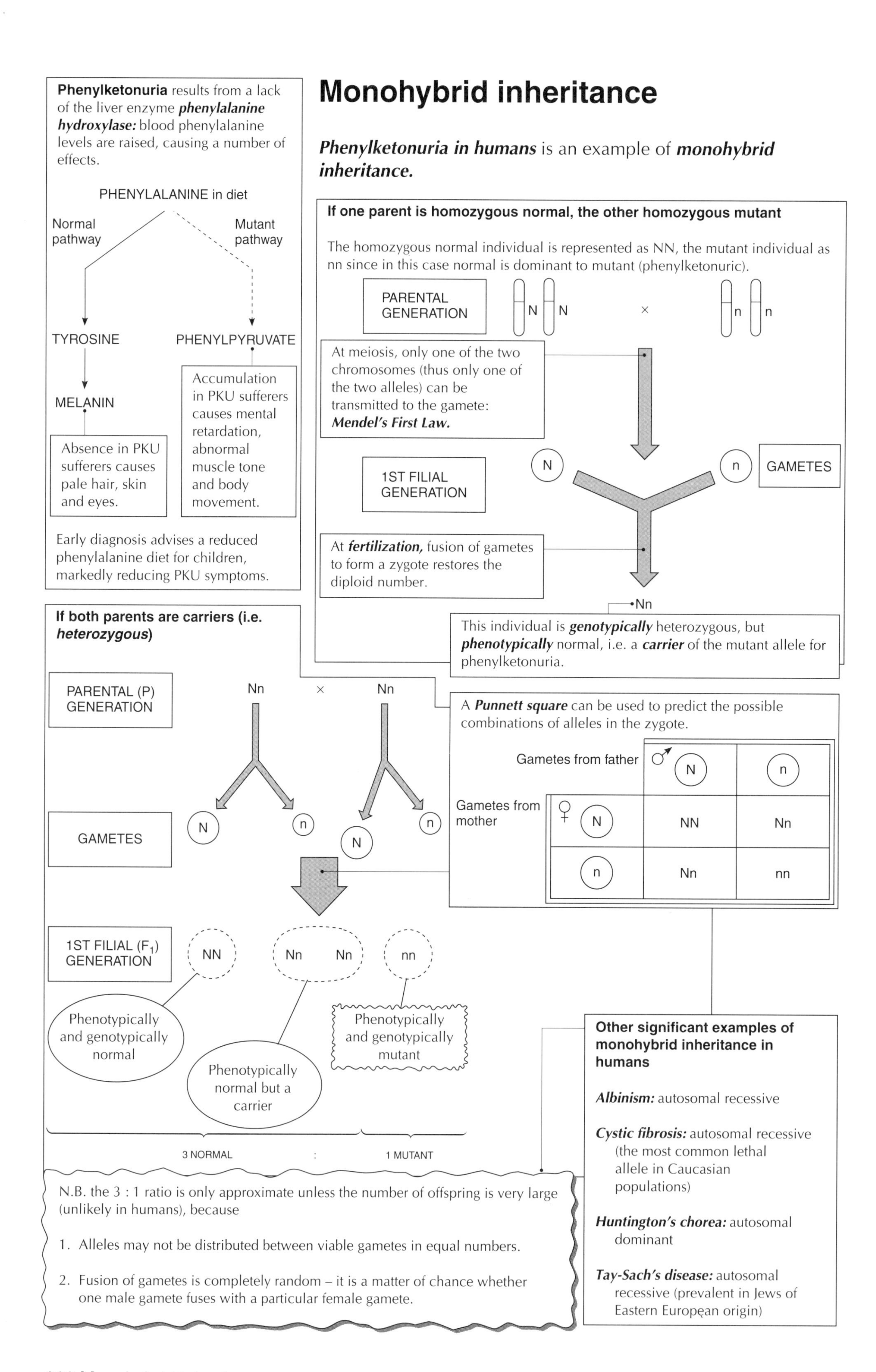
Monohybrid inheritance
Phenylketonuria in humans is an example of monohybrid inheritance.
Phenylketonuria results from a lack of the liver enzyme phenylalanine hydroxylase: blood phenylalanine levels are raised, causing a number of effects.
PHENYLALANINE in diet
Normal pathway
Mutant pathway
TYROSINE
PHENYLPYRUVATE
MELANIN
Absence in PKU sufferers causes pale hair, skin and eyes.
Accumulation in PKU sufferers causes mental retardation, abnormal muscle tone and body movement.
Early diagnosis advises a reduced phenylalanine diet for children, markedly reducing PKU symptoms.
If one parent is homozygous normal, the other homozygous mutant
The homozygous normal individual is represented as NN, the mutant individual as nn since in this case normal is dominant to mutant (phenylketonuric).
PARENTAL GENERATION
N N × n n
At meiosis, only one of the two chromosomes (thus only one of the two alleles) can be transmitted to the gamete: Mendel's First Law.
N
n
GAMETES
1ST FILIAL GENERATION
At fertilization, fusion of gametes to form a zygote restores the diploid number.
Nn
This individual is genotypically heterozygous, but phenotypically normal, i.e. a carrier of the mutant allele for phenylketonuria.
If both parents are carriers (i.e. heterozygous)
PARENTAL (P) GENERATION
Nn × Nn
GAMETES
N n N n
1ST FILIAL (F1) GENERATION
NN
Nn Nn
nn
Phenotypically and genotypically normal
Phenotypically normal but a carrier
Phenotypically and genotypically mutant
3 NORMAL : 1 MUTANT
A Punnett square can be used to predict the possible combinations of alleles in the zygote.
Gametes from father ♂ N n
Gametes from mother ♀
N NN Nn
n Nn nn
N.B. the 3 : 1 ratio is only approximate unless the number of offspring is very large (unlikely in humans), because
1. Alleles may not be distributed between viable gametes in equal numbers.
2. Fusion of gametes is completely random – it is a matter of chance whether one male gamete fuses with a particular female gamete.
Other significant examples of monohybrid inheritance in humans
Albinism: autosomal recessive
Cystic fibrosis: autosomal recessive (the most common lethal allele in Caucasian populations)
Huntington's chorea: autosomal dominant
Tay-Sach's disease: autosomal recessive (prevalent in Jews of Eastern European origin)

Linkage between genes prevents free recombination of alleles.

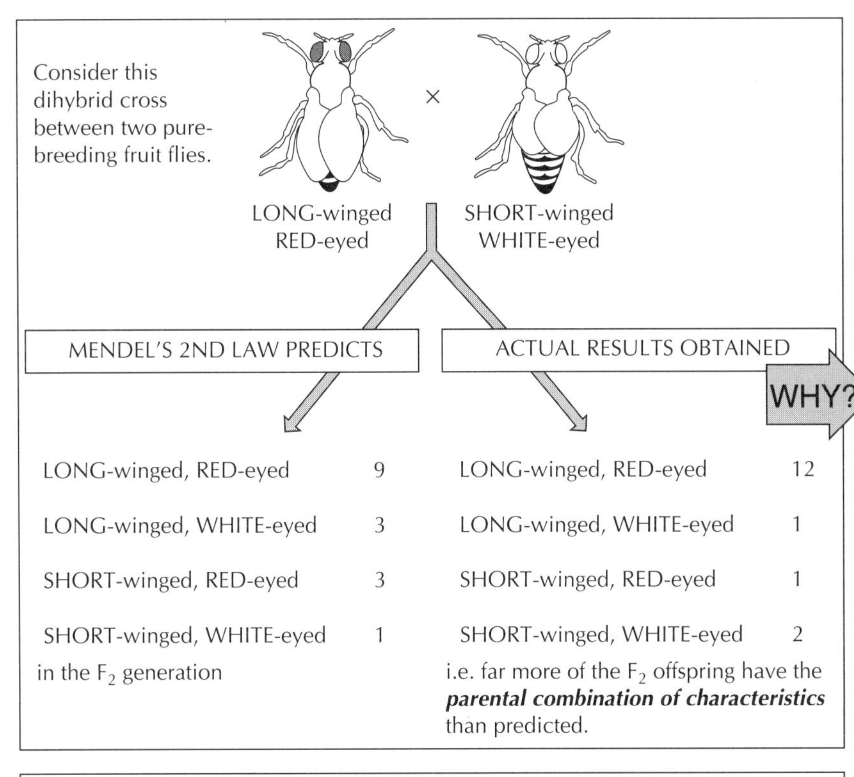

The genes for ***wing length*** and ***eye colour*** are ***linked.*** This means that ***they are located on the same chromosome*** and thus tend to ***pass into gametes together.***

i.e. ***parental*** chromosomes can be represented as

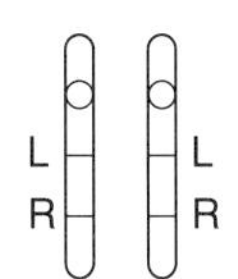

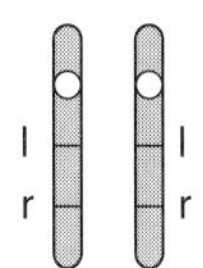

producing gametes

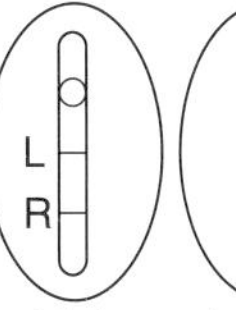

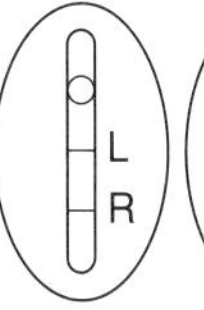

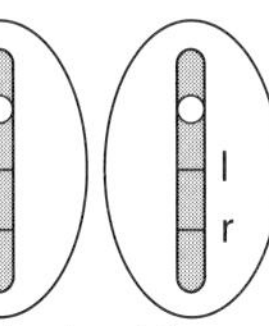

which produce F_1 individuals with

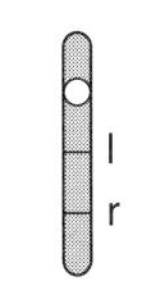

and since LR are linked, as are lr, the F_2 offspring will tend to be

from ♀ \ from ♂	LR	lr
LR	LONG, RED	LONG, RED
lr	LONG, RED	SHORT, WHITE

That is, the alleles tend to remain in the original parental combinations and so parental phenotypes predominate.

How can linked alleles be separated?

During ***meiosis,*** homologous chromosomes pair up to form ***bivalents*** and replicate to form ***tetrads*** of chromatids.

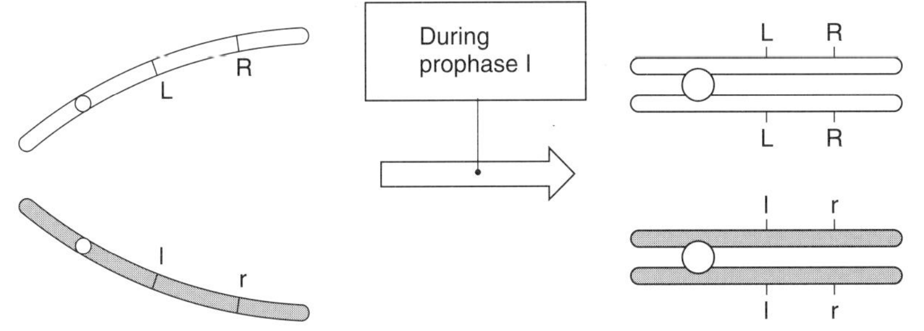

As the chromatids lie alongside one another it is possible for ***crossing over*** (the exchange of genetic material between adjacent members of a homologous pair) to occur.

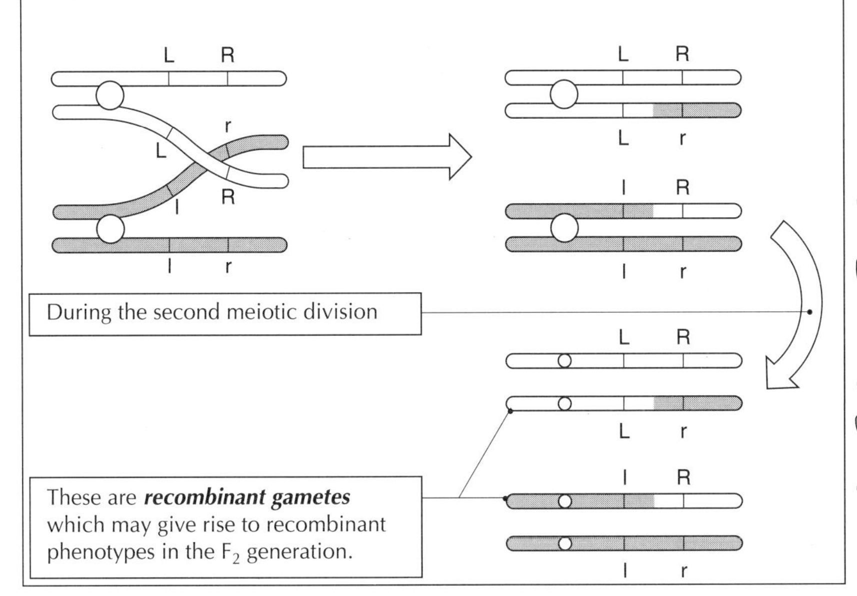

One well-known example of ***linkage in humans*** involves the genes for ***ABO blood groups*** and the ***nail-patella syndrome.***

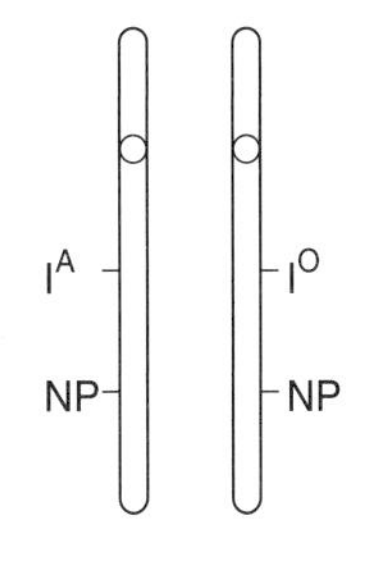

The NP allele is a dominant one: people with this syndrome have small, discoloured nails, and the patella is missing or small and pushed to one side.

Most people with N-P syndrome belong to either A or O blood group.

Sex linkage and the inheritance of sex

Any genes on the X chromosomes will be inherited by both sexes, but whereas the male can only receive ***one*** of the alleles (he will be XY, and therefore must be ***homozygous*** for the X-linked allele) the female will be XX and thus may be either ***homozygous*** or ***heterozygous.*** This gives females a tremendous genetic advantage since any recessive lethal allele will not be expressed in the heterozygote.

For example, ***haemophilia*** is an X-linked condition.

Normal gene = H, mutant gene = h

PARENTS X^HX^h female, carrier × X^HY male, normal

i.e. ***both parents have normal phenotype …***

GAMETES X^H X^h X^H Y

F_1 OFFSPRING

♂ GAMETES / ♀ GAMETES	X^H	Y
X^H	X^HX^H	X^HY
X^h	X^HX^h	X^hY

ie. normal, female X^HX^H; carrier, female X^HX^h; normal, male X^HY; haemophiliac, male X^hY

… but may have a haemophiliac son.

Other significant X-linked conditions include ***Duchenne muscular dystrophy, red-green colour blindness*** and ***coat colour in cats*** (where the alleles for ginger (G) and black (g) produce tortoiseshell in the heterozygote X^GX^g: thus there should, in theory, be ***no male tortoiseshell cats!***

Any genes carried on the Y chromosome will be received by ***all*** the male offspring – there is little space for other than sex genes, but one well-known example concerns ***webbed toes.***

GAMETES $X\ Y^W$ (only Y chromosome carries W gene) × X X

OFFSPRING $X\ Y^W$ $X\ Y^W$ (males with webbed toes) X X X X (normal females)

In mammals sex is determined by two chromosomes which are very different to one another. These are the ***heterosomes*** – the male is ***heterogametic*** (XY: can produce both X and Y gametes) and the female is ***homogametic*** (can only produce X gametes).

♀ (XX) ♂ (XY)

These sections are ***homologous*** and carry no genes of sex determination.

These ***non-homologous*** sections carry the genes concerned with sex determination but are of sufficient size to carry other genes. Such genes are ***sex-linked.***

Inheritance of sex is a special form of Mendelian segregation.

♂ XY × ♀ XX

GAMETES X Y X X

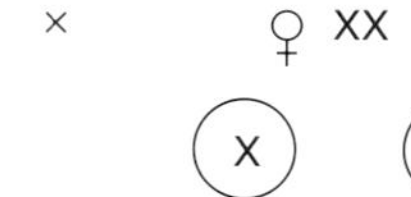

F_1 generation: sex of offspring can be determined from a Punnett square.

♂ GAMETES / ♀ GAMETES	X	Y
X	XX (female)	XY (male)
X	XX (female)	XY (male)

Theoretically there should be a 1 : 1 ratio of male : female offspring. In humans various factors can upset the ratio – the Y sperm tend to have greater mobility; the XY zygote and embryo is more delicate than the XX embryo. The balance is just about maintained.

Dihybrid inheritance involves the transmission of ***two pairs of alleles*** at the same time but independently of one another.

Gregor Mendel was an Austrian monk who studied patterns of inheritance in the garden pea. He made several proposals concerning these patterns: these proposals have been found to hold true for many other organisms. One organism which has been widely used for experiments on inheritance is the fruit fly, *Drosophila melanogaster*, which has a number of advantages for such studies:

1. It has a rapid generation time (10 days) and produces many offspring from each mating so that statistical analysis can be applied to results.
2. It is easily cultured in small, convenient containers (milk bottles!) on a simple growth medium.
3. Males and females are easily distinguished so that controlled matings are possible.
4. The flies have a number of obvious external characteristics which are easily mutated. These include wing length, body colour and eye shape.

Flies which differ in ***two pairs of characteristics*** (e.g. ***body colour*** and ***wing shape)*** may be mated.

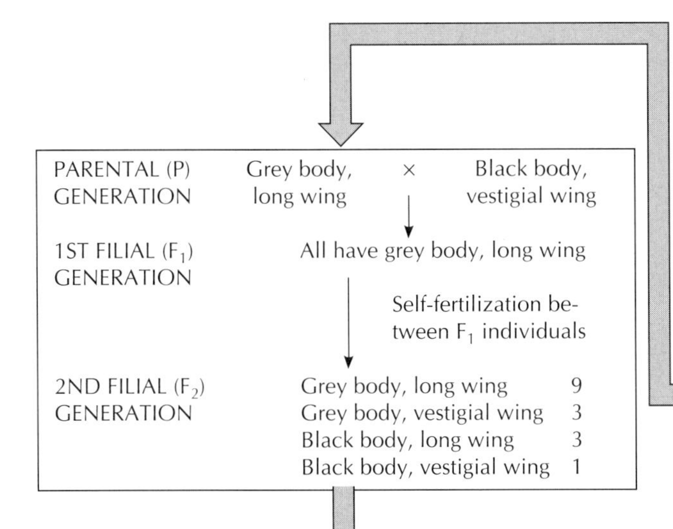

A phenotypic ratio of 9:3:3:1 would seem to be complex, but Mendel explained this as ***two separate monohybrid crosses*** (i.e. ***grey v. black*** and ***long v. vestigial) occurring at the same time.***

ie. GREY v. BLACK = (9+3) : (3+1)
= 12 : 4
= 3 : 1

LONG v. VESTIGIAL = (9+3) : (3+1)
= 12 : 4
= 3 : 1

i.e. 9 : 3 : 3 : 1 is the same as 3 : 1 x 3 : 1

Thus the inheritance of ***body colour*** had not influenced the inheritance of ***wing shape.***

Mendel's Second Law (the law of independent assortment)

'Each member of a pair of alleles may combine randomly with either of another pair'

In this example, the allele for ***grey*** body may combine equally often with the allele for ***long*** wing or with the allele for ***vestigial*** wing.

Using genetic symbols

Let G = grey, g = black, L = long, l = vestigial

P GGLL × ggll

Gametes (GL) (gl)

F_1 GgLl

Gametes (GL)  (Gl) (gL)  (gl)

These will be produced in equal numbers, according to Mendel's Second Law.

F_2 The possible combinations of gametes are most easily derived using a ***Punnett square.***

♂ GAMETES / ♀ GAMETES	GL	Gl	gL	gl
GL	GGLL	GGLl	GgLL	GgLl
Gl	GGLl	GGll	GgLl	Ggll
gL	GgLl	GgLl	ggLL	ggLl
gl	GgLl	Ggll	ggLl	ggll

or, phenotpically
9 GREY BODY, LONG WING (both G and L in zygote)
3 GREY BODY, VESTIGIAL WING (G and ll in zygote)
3 BLACK BODY, LONG WING (gg and L in zygote)
1 BLACK BODY, VESTIGIAL WING (ggll in zygote)

Variation is the basis of evolution.

Origins of variation

May be ***non-heritable*** (e.g. sunburn in a light-skinned individual) or ***heritable*** (e.g. skin colour in different races). The second type, which result from genetic changes, are the most significant in evolution.

Mutation is any change in the structure or the amount of DNA in an organism.

- A ***gene*** or ***point mutation*** occurs at a single locus on a chromosome – most commonly by ***deletion, addition*** or ***substitution*** of a nucleotide base. Examples are sickle cell anaemia, phenylketonuria and cystic fibrosis.
- A ***change in chromosome structure*** occurs when a substantial portion of a chromosome is altered. For example, Cri-du-chat syndrome results from ***deletion*** of a part of human chromosome 5, and a form of white blood cell cancer follows ***translocation*** of a portion of chromosome 8 to chromosome 14.
- ***Aneuploidy*** (typically the ***loss or gain of a single chromosome***) results from ***non-disjunction*** in which chromosomes fail to separate at anaphase of meiosis. The best known examples are Down's syndrome (extra chromosome 21), Klinefelter's syndrome (male with extra X chromosome) and Turner's syndrome (female with one fewer X chromosome).
- ***Polyploidy*** (the presence of additional ***whole sets of chromosomes***) most commonly occurs when one or both gametes is diploid, forming a polyploid on fertilization. Polyploidy is rare in animals, but there are many important examples in plants, e.g. bananas are triploid, and tetraploid tomatoes are larger and richer in vitamin C.

Sexual or genetic recombination is a most potent force in evolution, since it reshuffles genes into new combinations. It may involve

free assortment in gamete formation
crossing over during meiosis
random fusion during zygote formation.

Discontinuous variation occurs when a characteristic is either present or absent (the two extremes) and there are no intermediate forms.

Such variations do not give normal distribution curves but bar charts are often used to illustrate the distribution of a particular characteristic in a population.

NUMBER IN GROUP
O A B AB
BLOOD GROUP

Examples are human blood groups in the ABO system (O, A, B or AB), basic fingerprint forms (loop, whorl or arch) and tongue-rolling (can or cannot).

A characteristic which shows discontinuous variation is normally controlled by a single gene – there may be two or more alleles of this gene.

Continuous variation occurs when there is a gradation between one extreme and the other of some given characteristic – all individuals exhibit the characteristic but to differing extents.

If a frequency distribution is plotted for such a characteristic a ***normal*** or ***Gaussian distribution*** is obtained.

NUMBER IN GROUP
150 155 160 165 170 175 180 185
SIZE/cm

The ***mean*** is the average number of such a group (i.e. the total number of individuals divided by the number of groups), the ***mode*** is the most common of the groups and the ***median*** is the central value of a set of values.

Typical examples are height, mass, handspan, or number of leaves on a plant.

Characteristics which show continuous variation are controlled by the combined effect of a number of genes, called ***polygenes,*** and are therefore ***polygenic characteristics.***

Natural selection may be a potent force in ***evolution.***

Much variation is of the ***continuous type*** , i.e. a range of phenotypes exists between two extremes. The range of phenotypes within the environment will typically show a ***normal distribution.***

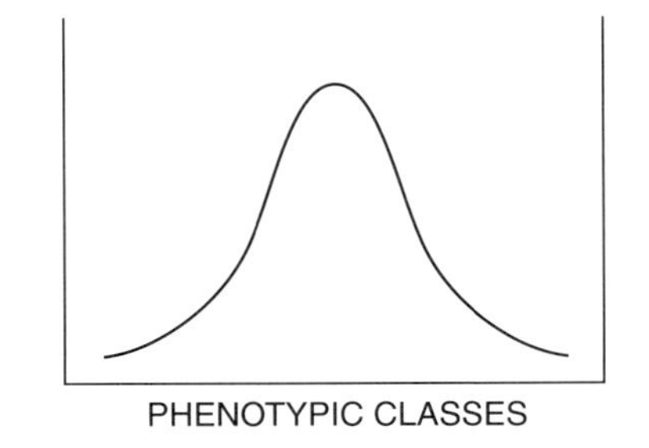

The modern ***neo-Darwinian*** theory accepts that:

1. Some harmful alleles may survive, but the reproductive potential of the individual possessing such an allele will be reduced.
2. Selective advantages and disadvantages of an allele relate to one environment at one particular time, i.e. an allele does not always contribute to 'fitness' but only under certain conditions.

Plants and animals in Nature produce more offspring than can possibly survive, yet the population remains relatively constant. There must be many deaths in Nature.

Overproduction of this type leads to ***competition*** – for food, shelter and breeding sites, for example. There is thus a ***struggle for existence.*** Those factors in the environment for which competition occurs represent ***selection pressures.***

Within a population of individuals there may be considerable

Variation means that some individuals possess characteristics which would be advantageous in the struggle for existence (and some would be the opposite, of course).

Those possessing the best combination of characteristics would be more competitive in the struggle for existence: they would be more 'fit' to cope with the selection pressures imposed by the environment. This is ***natural selection*** and promotes ***survival of the fittest.***

If variation is ***heritable*** (i.e. caused by an alteration in genotype) new generations will tend to contain a higher proportion of individuals suited to survival.

Stabilizing selection favours intermediate phenotypic classes and operates against extreme forms – there is thus a ***decrease*** in the frequency of alleles representing the extreme forms.

Stabilizing selection operates when the phenotype corresponds with optimal environmental conditions, and competition is not severe. It is probable that this form of selection has favoured heterozygotes for ***sickle cell anaemia*** in an environment in which ***malaria*** is common, and also works against ***extremes of birth weight*** in humans.

Directional selection favours one phenotype at one extreme of the range of variation. It moves the phenotype towards a new optimum environment; then stabilizing selection takes over. There is a change in the allele frequencies corresponding to the new phenotype.

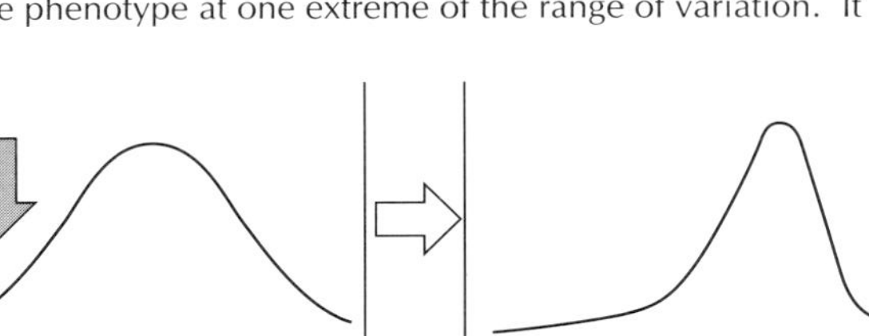

Directional selection has occurred in the case of the peppered moth, *Biston betularia*, where the dark form was favoured in the sooty suburban environments of Britain during the industrial revolution: ***industrial melanism.*** Another significant example is the development of ***antibiotic resistance*** in populations of bacteria – mutant genes confer an advantage in the presence of an antibiotic.

Disruptive selection is the rarest form of selection and is associated with a variety of selection pressures operating within one environment.

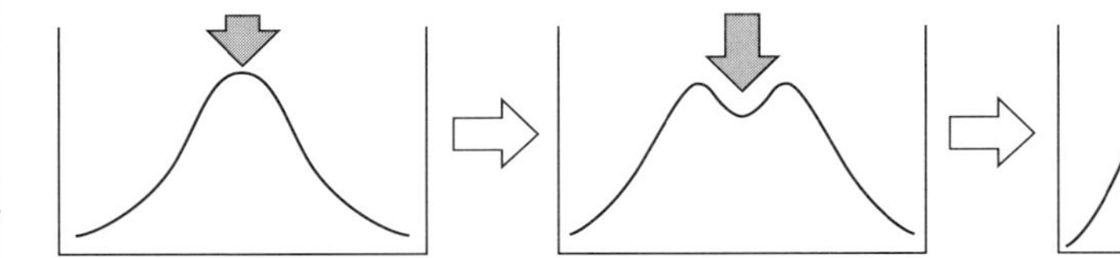

This form of selection promotes the co-existence of more than one phenotype, the condition of ***polymorphism (balanced*** polymorphism when no one selective agent is more important than any other). Important examples are:

1. ***Colour*** (yellow/brown) and ***banding pattern*** (from 0–5) in *Cepaea nemoralis.*
2. ***Three phenotypes*** (corresponding to $HbHb/HbHb^s/Hb^sHb^s$) show an uneven distribution of the sickle-cell allele in different areas of the world.

Artificial selection

Artificial selection occurs when humans, rather than environmental factors, determine which genotypes will pass to successive generations.

POLYPLOIDY AND PLANT BREEDING

Polyploids contain ***multiple sets of chromosomes*** (chromosome multiplication can be induced by treatment with ***colchicine*** during mitosis – this inhibits spindle formation and prevents chromatid separation).

Autopolyploids (all chromosomes from the ***same*** species) e.g. all ***bananas*** are ***triploid*** – they are infertile and contain no seeds. Most ***potatoes*** are ***tetraploid*** – cells are bigger and tubers are larger. Cultivated ***strawberries*** are ***octoploid.***

Allopolyploidy (sets of chromosomes from ***different*** species) is possible if the two species have a chromosome complement similar in number and shape. This might allow plant breeders to ***combine the beneficial characteristics of more than one species.***

The evolution of ***bread wheat*** is an important example.

Wild wheat: has brittle ears which fall off on harvesting.

↓ SELECTIVE BREEDING

Einkorn wheat: non-brittle but low yielding.

↓ POLYPLOIDY with *Agropyron* grass

Emmer wheat: high yielding but difficult to separate seed during threshing.

↓ POLYPLOIDY with *Aegilops* grass

Bread wheat: high yielding with easily separated 'naked' seeds.

INBREEDING AND OUTBREEDING

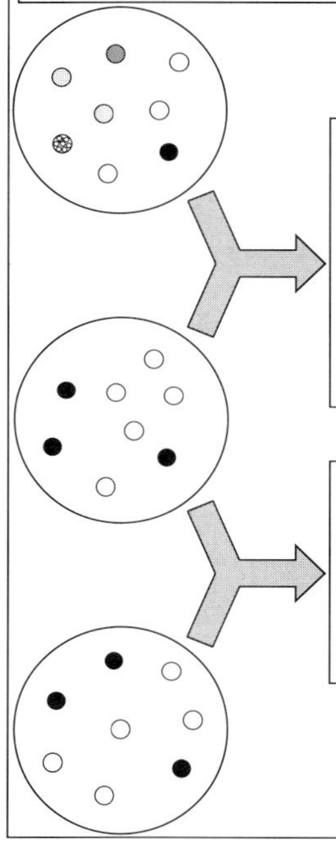

Outbreeding occurs when there is selective controlled reproduction between members of genetically distant populations (different strains or even, for plants, closely related but different species).

Tends to ***introduce new and superior phenotypes:*** the progeny are known as ***hybrids*** and the development of improved characteristics is called ***heterosis*** or ***hybrid vigour.***

e.g. introduction of disease resistance from wild sheep to domestic strains;
combination of shorter-stemmed 'wild' wheat and heavy yielding 'domestic' wheat.

This may result from increased numbers of dominant alleles or from new opportunities for gene interaction.

Inbreeding occurs when there is selective reproduction between closely related individuals, e.g. between offspring of same litter or between parent and child.

Tends to ***maintain desirable characteristics***

e.g. uniform height in maize (easier mechanical harvesting);
maximum oil content of linseed (more economical extraction);
milk production by Jersey cows (high cream content).

But it may cause ***reduced fertility*** and ***lowered disease resistance*** as genetic variation is reduced.

Thus inbreeding is not favoured by animal breeders.

PROTOPLAST FUSION

PROTOPLAST FUSION is a modern method for production of hybrids in plants.

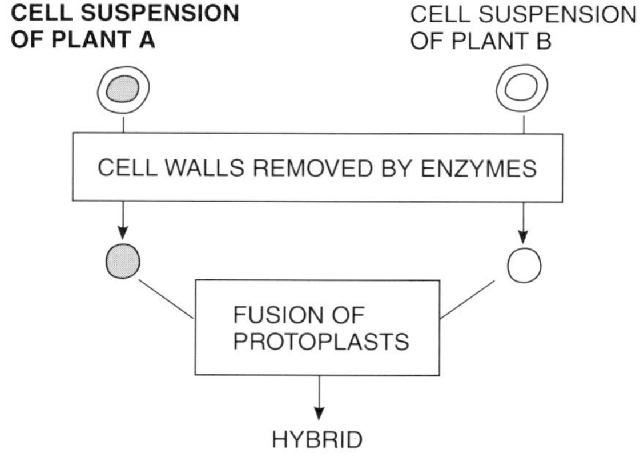

CELL SUSPENSION OF PLANT A — CELL SUSPENSION OF PLANT B

CELL WALLS REMOVED BY ENZYMES

FUSION OF PROTOPLASTS

HYBRID

e.g. production of virus-resistant tobacco.

TECHNIQUES WITH ANIMALS

TECHNIQUES WITH ANIMALS are less well advanced than those with plants because:

a animals have a longer generation time and few offspring
b more food will be made available from improved plants
c there are many ethical problems which limit genetic experiments with animals.

Two important animal techniques are:

Artificial insemination: allows sperm from a male with desirable characteristics to fertilize a number of female animals.

Embryo transplantation: allows the use of ***surrogate mothers*** (thus increasing number of offspring) and ***cloning*** (production of many identical animals with the desired characteristics).

Reproductive isolation and speciation

Reproductive isolation is essential for speciation: ***allopatric speciation*** occurs when populations occupy different environments; ***sympatric speciation*** occurs when populations are reproductively isolated within the same environment.

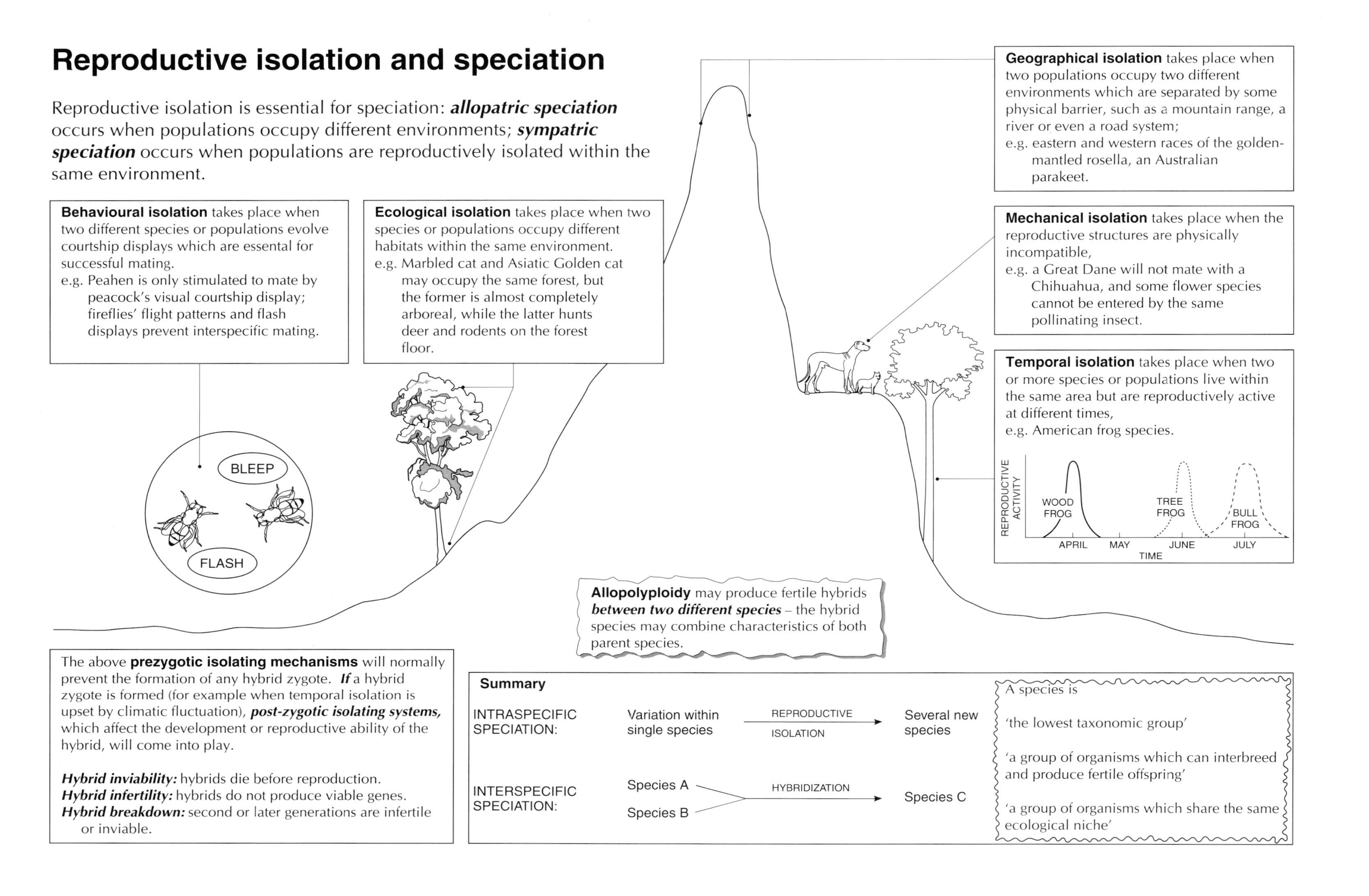

Behavioural isolation takes place when two different species or populations evolve courtship displays which are essential for successful mating.
e.g. Peahen is only stimulated to mate by peacock's visual courtship display; fireflies' flight patterns and flash displays prevent interspecific mating.

Ecological isolation takes place when two species or populations occupy different habitats within the same environment.
e.g. Marbled cat and Asiatic Golden cat may occupy the same forest, but the former is almost completely arboreal, while the latter hunts deer and rodents on the forest floor.

Geographical isolation takes place when two populations occupy two different environments which are separated by some physical barrier, such as a mountain range, a river or even a road system;
e.g. eastern and western races of the golden-mantled rosella, an Australian parakeet.

Mechanical isolation takes place when the reproductive structures are physically incompatible,
e.g. a Great Dane will not mate with a Chihuahua, and some flower species cannot be entered by the same pollinating insect.

Temporal isolation takes place when two or more species or populations live within the same area but are reproductively active at different times,
e.g. American frog species.

Allopolyploidy may produce fertile hybrids ***between two different species*** – the hybrid species may combine characteristics of both parent species.

The above **prezygotic isolating mechanisms** will normally prevent the formation of any hybrid zygote. ***If*** a hybrid zygote is formed (for example when temporal isolation is upset by climatic fluctuation), ***post-zygotic isolating systems,*** which affect the development or reproductive ability of the hybrid, will come into play.

Hybrid inviability: hybrids die before reproduction.
Hybrid infertility: hybrids do not produce viable genes.
Hybrid breakdown: second or later generations are infertile or inviable.

Summary

INTRASPECIFIC SPECIATION: Variation within single species —REPRODUCTIVE ISOLATION→ Several new species

INTERSPECIFIC SPECIATION: Species A + Species B —HYBRIDIZATION→ Species C

A species is

'the lowest taxonomic group'

'a group of organisms which can interbreed and produce fertile offspring'

'a group of organisms which share the same ecological niche'

Gene cloning

involves ***recombination***, ***transformation*** and ***selection***.

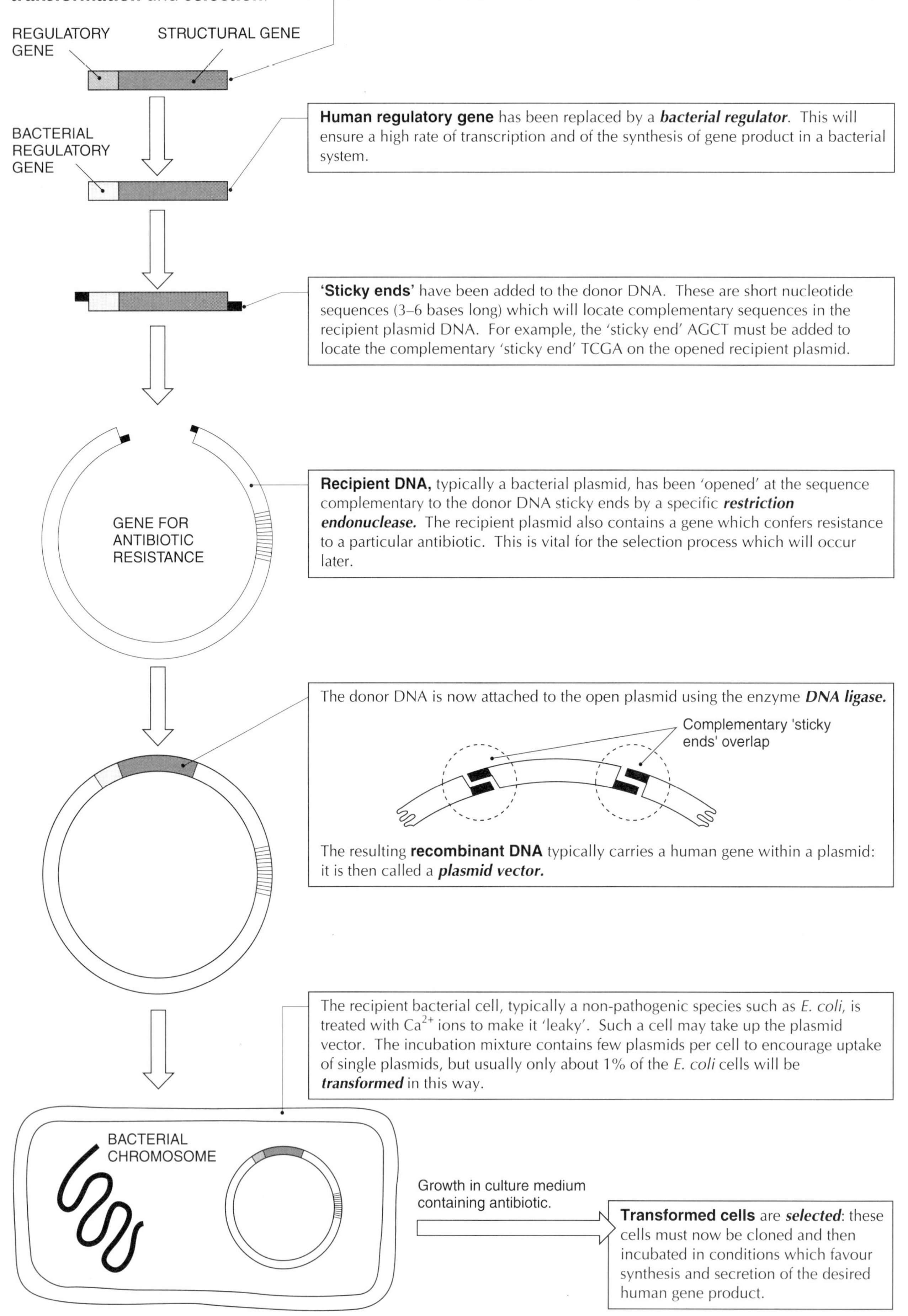

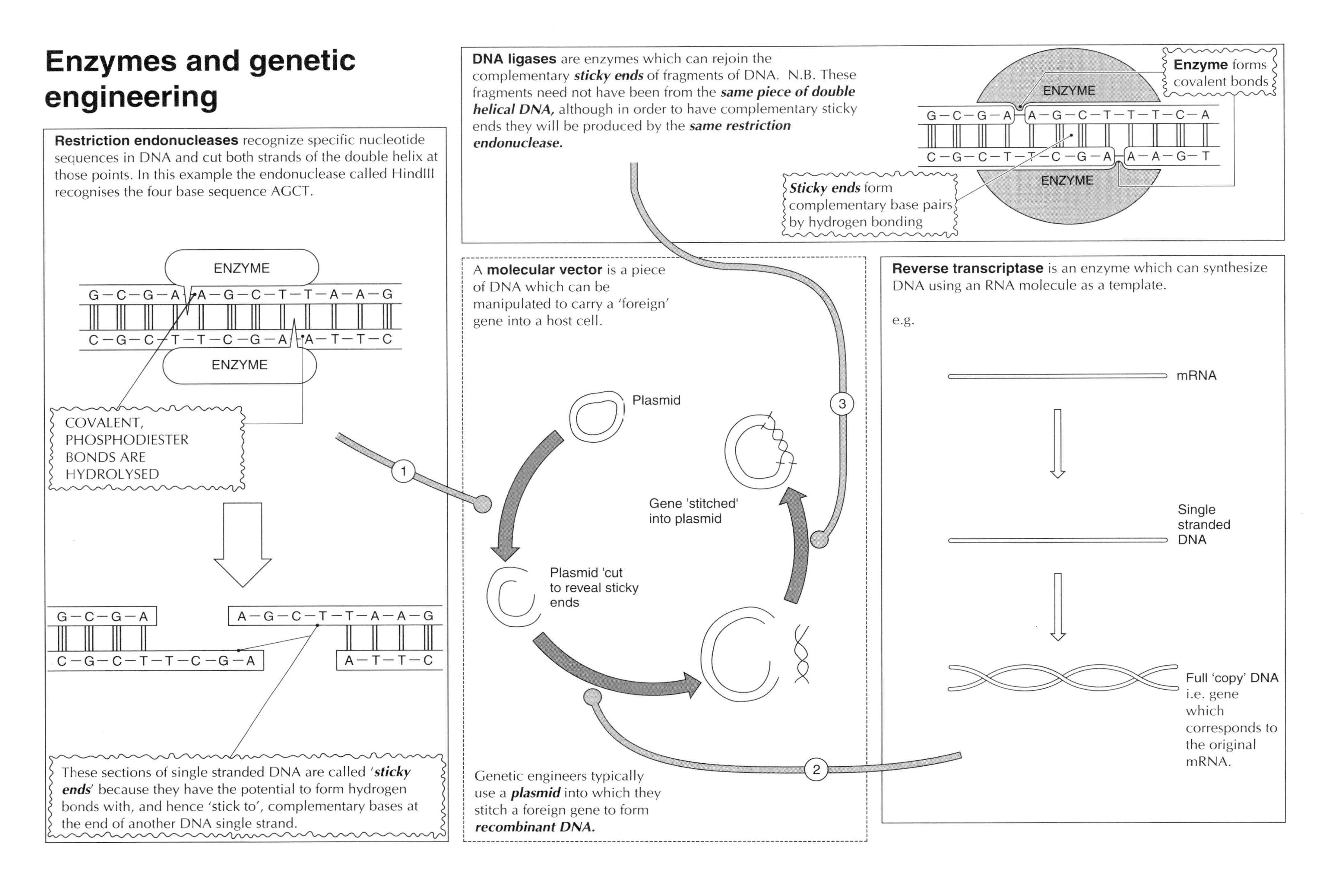
Enzymes and genetic engineering
Restriction endonucleases recognize specific nucleotide sequences in DNA and cut both strands of the double helix at those points. In this example the endonuclease called HindIII recognises the four base sequence AGCT.
ENZYME
G–C–G–A A–G–C–T–T–A–A–G
C–G–C–T–T–C–G–A A–T–T–C
ENZYME
COVALENT, PHOSPHODIESTER BONDS ARE HYDROLYSED
G–C–G–A
C–G–C–T–T–C–G–A
A–G–C–T–T–A–A–G
A–T–T–C
These sections of single stranded DNA are called 'sticky ends' because they have the potential to form hydrogen bonds with, and hence 'stick to', complementary bases at the end of another DNA single strand.
DNA ligases are enzymes which can rejoin the complementary sticky ends of fragments of DNA. N.B. These fragments need not have been from the same piece of double helical DNA, although in order to have complementary sticky ends they will be produced by the same restriction endonuclease.
ENZYME
G–C–G–A A–G–C–T–T–T–C–A
C–G–C–T–T–C–G–A A–A–G–T
ENZYME
Enzyme forms covalent bonds
Sticky ends form complementary base pairs by hydrogen bonding
A molecular vector is a piece of DNA which can be manipulated to carry a 'foreign' gene into a host cell.
Plasmid
Plasmid 'cut to reveal sticky ends
Gene 'stitched' into plasmid
1
2
3
Genetic engineers typically use a plasmid into which they stitch a foreign gene to form recombinant DNA.
Reverse transcriptase is an enzyme which can synthesize DNA using an RNA molecule as a template.
e.g.
mRNA
Single stranded DNA
Full 'copy' DNA i.e. gene which corresponds to the original mRNA.

INDEX